THE THEORY OF EVERYTHING

THE QUEST TO EXPLAIN ALL REALITY

Don Lincoln, Ph.D.

THE GREAT COURSES®

PUBLISHED BY:

THE GREAT COURSES
Corporate Headquarters
4840 Westfields Boulevard, Suite 500
Chantilly, Virginia 20151-2299
Phone: 1-800-832-2412
Fax: 703-378-3819
www.thegreatcourses.com

Printed in the United States of America

FSC
www.fsc.org
MIX
Paper from responsible sources
FSC® C011935

RAINFOREST ALLIANCE CERTIFIED

DON LINCOLN, PH.D.

SENIOR SCIENTIST
FERMI NATIONAL ACCELERATOR LABORATORY

Don Lincoln is a Senior Scientist at Fermi National Accelerator Laboratory. His research time has been divided between studying data from the Tevatron Collider (until it closed in 2011) and data from the CERN Large Hadron Collider, located outside of Geneva, Switzerland. Dr. Lincoln is also a Guest Professor of High Energy Physics at the University of Notre Dame. He received his Ph.D. in Experimental Particle Physics from Rice University.

Dr. Lincoln is the coauthor of more than 1000 scientific publications that range over subjects from microscopic black holes and extra dimensions to the elusive Higgs boson. His most noteworthy scientific accomplishments include being part of the teams that discovered the top quark in 1995 and confirmed the Higgs boson in 2012. Dr. Lincoln is interested in everything about particle physics and cosmology, but his most burning interest is in understanding the reasons for why there are so many known subatomic building blocks. He searches for possible constituents of the quarks and leptons, which are treated in the standard model of particle physics as featureless and point-like particles.

When Dr. Lincoln isn't exploring the energy frontier, he enjoys communicating the excitement of his and his colleagues' cutting-edge research with the public. He has authored 3 books for the public about particle physics that have been translated into Polish, Russian, German, Korean, Portuguese, and Chinese: *Understanding the Universe: From Quarks to the Cosmos*; *The Quantum Frontier: The Large Hadron Collider*; and *The Large Hadron Collider: The Extraordinary Story of the Higgs Boson and Other Stuff That Will Blow Your Mind*. His fourth book, *Alien Universe: Extraterrestrial Life in Our Minds and in the Cosmos*, combines astrobiology and popular reports of alien visitation to weave together a complete tale of the possibility of life from other planets.

Dr. Lincoln has published many articles in periodicals that include *Analog Science Fiction and Fact*, *The Physics Teacher*, and *Scientific American*. His online articles have appeared on CNN.com, *The Huffington Post*, and Live Science. Dr. Lincoln's science outreach efforts resulted in him being awarded the 2013 Outreach Prize from the High Energy Physics Division of the European Physical Society. He is also a fellow of the American Physical Society and the American Association for the Advancement of Science.

Dr. Lincoln has given hundreds of lectures on 4 continents and to a broad range of audiences, but his favorite kind of audience is nonscientists who are interested in understanding how the world works. He has a well-regarded series of YouTube videos that explain frontier physics to a lay audience, and he is a blogger for the website of the popular television series *NOVA*. Dr. Lincoln also writes a weekly column for the online periodical *Fermilab Today*, which popularizes research papers as they are released.

You can follow Dr. Lincoln at http://www.facebook.com/Dr.Don.Lincoln. ■

TABLE OF CONTENTS

Introduction

Lecture Guides

Supplemental Material

THE THEORY OF EVERYTHING

THE QUEST TO EXPLAIN ALL REALITY

For as long as humanity has kept records, we have asked some very big questions: How did we get here? How did the universe come into existence? Does the universe have to be the way that it is?

While these questions were once exclusively the province of philosophy and religion, to determine quantifiable answers, we must turn to the realm of science. For more than 4 centuries, physicists have led the scientific quest toward the most fundamental answers to those questions. Their goal is to create a single theory, using a small number of building blocks and a few guiding principles, that will allow them to answer these questions as comprehensively, yet also as succinctly, as possible. The name that physicists use for this is a theory of everything.

In this course, you will learn to view the major advances in physics as a series of unifications. Isaac Newton unified the motion of a falling apple with the passage of the Moon around the Earth. A series of 19th-century scientists, culminating with James Clerk Maxwell, unified electricity and magnetism into a single force called electromagnetism. With the discovery of radiation in the early 20th century, scientists discovered the strong nuclear force, which holds together nuclei and the weak force, which is responsible for some forms of radioactivity. This course also describes the unification in the 1960s of the electromagnetic and weak forces into a combined and more fundamental electroweak force.

These achievements have allowed current-day physicists to categorize the phenomena of the cosmos into 2 broad categories: a theory of gravity and a preliminary and incomplete grand unified theory of everything else. The course begins with a deep dive into the latter—into past and prospective steps toward a complete grand unified theory, excluding gravity, as currently expressed in the standard model.

The standard model of particle physics is our modern understanding of the subatomic realm and combines both quantum mechanics and special relativity to create quantum field theories. It explains the nature and behavior of matter as being caused by 2 classes of fundamental building blocks (called quarks and leptons), governed by 3 forces (electromagnetism and the strong and weak nuclear forces). Each of these topics is covered in at least 1 lecture. In addition, the recently discovered Higgs boson provides a capstone for the standard model.

But the standard model alone does not provide a full grand unified theory of the strong, weak, and electromagnetic forces, so the course also addresses particle physics beyond the standard model. Subatomic particles called neutrinos have been observed to change their identity, morphing from one type of neutrino into another before converting back again.

Devising a theory of everything also requires inventing new and powerful models that bring together disparate ideas, such as supersymmetry and the search for a possible level of structure even below the level of quarks and leptons. Mathematical symmetries have played a fundamental role in generating these and other new theories.

Next, the course turns to gravity. After a brief introduction to Newton's contributions, the course gives an overview of Einstein's theory of general relativity. Once you have an appreciation of those well-understood topics, the course dives into efforts to unify gravity and the standard model, with descriptions of superstrings and quantum gravity. Quantum gravity is on the forefront of modern theoretical research, and even top-notch scientists don't know what the final answer will be.

While both the standard model and general relativity are incredible intellectual triumphs, neither one explains enough to be considered a theory of everything—nor are they compatible with one another. To make further progress, researchers look at phenomena and measurements that are not described by either theory for clues as to what kinds of changes are necessary to make improved theories.

In the realm of gravity, there is both dark matter and dark energy, neither of which can be accommodated by existing models. Dark matter is hypothesized to be some kind of particulate matter that persists throughout the universe and explains why galaxies rotate faster than can be explained by the observed matter and current theory of gravity. In contrast, dark energy is an energy field that permeates the entire cosmos and is invoked to explain why the expansion of the universe is accelerating. The possibility of extra dimensions has been proposed to account for the extreme weakness of gravity compared to the other fundamental forces.

In the quantum world, scientists do not understand why our universe is made solely of matter when our best theories suggest that it should consist equally of matter and antimatter. In addition, a theory of everything must not only explain what we see in our universe, but also how our universe began and evolved. The big bang theory explains a lot, but it is silent on the moment of creation itself. The course covers both what is known and also speculation about the earliest moments of the universe. The course even covers the humbling concept of the multiverse, the idea that our universe is but one of many and that other universes may obey laws of physics that are quite different than our familiar ones.

This course dives deep into what we know, what we don't know, and how we hope to move forward to a unified understanding of everything.

1 TWO PROTOTYPE THEORIES OF EVERYTHING

The ultimate goal of science is an understanding of the fundamental rules of the universe. Over the centuries, scientists have combined their thinking and observations and distilled them into 2 distinct theories: Einstein's general theory of relativity, which is our best description of how gravity works, and the standard model of particle physics, which covers everything else. The 2 theories are incompatible, however, and we need to invent a theory of quantum gravity and merge it into the standard model.

CHEMISTRY AS A BASIS FOR EVERYTHING

- In the 18th and early 19th centuries, a legion of early chemists discovered that there were several fundamental substances they called elements that seemed to combine in different configurations. Chemistry occurred when the elements combined or broke apart.

- Another key clue was revealed in the early 1800s, when English scientist John Dalton discovered that when elements combined, they combined in fixed ratios. Essentially, his work verified that atoms existed.

JOHN DALTON
1766–1844

- People had talked about atoms for more than 2000 years, but it took until the first years of the 19th century that we had experimental data that verified the idea. And with the discovery of atoms, our understanding of the behavior of matter had one of its first unification moments: All matter was made of atoms of elements.

- Having this new unit of measure also made it possible to begin organizing even more observations. There were highly reactive elements and ones that were less reactive; some elements liked to interact with other elements more than others. And some mixed in a few ways—for example, carbon monoxide, which consists of a carbon and an oxygen atom, and carbon dioxide, which consists of a carbon and 2 oxygen atoms.

- In 1869, many additional observations were brought together when Russian chemist Dmitri Mendeleev invented what is now called the periodic table of elements. He took the discrete pieces of knowledge developed by a century of chemists and alchemists and organized the elements in a big grid, in which each column contains atoms with similar chemical properties.

DMITRI MENDELEEV
1834–1907

- The periodic table is a pretty amazing achievement, but perhaps even more amazing is that when Mendeleev started to work out the table, he didn't know about the more than 100 elements that are on the table today. He only knew about slightly more than 50, which means that he had some holes in his table, but he was able to work out the patterns with incomplete information.

- When Mendeleev wrote it down, he didn't know why the elements had the properties they do and why the patterns repeated. In fact, answering those questions would take another 50 years, when we finally understood the nature of the atom and the laws of quantum mechanics. All he knew at the time was that some elements reacted similarly and that the mass of these similar elements was different. But those 2 observations led him to his key insight.

- There was a point in the 1880s when the best scientific minds in the world would look at the periodic table and wonder what those patterns were telling them. It would take the unifying ideas of nuclear and atomic physics to explain the mysteries of chemistry.

MENDELEEV'S PERIODIC TABLE

PERIOD

GROUP	I	II	III	IV	V	VI	VII	VIII
1	H = 1							
2	Li = 7	Be = 9.4	B = 11	C = 12	N = 14	O = 16	F = 19	
3	Na = 23	Mg = 24	Al = 27.3	Si = 28	P = 31	S = 32	Cl = 35.5	
4	K = 39	Ca = 40	? = 44	Ti = 48	V = 51	Cr = 52	Mn = 55	Fe = 56, Co = 59, Ni = 59
5	Cu = 63	Zn = 65	? = 68	? = 72	As = 75	Se = 78	Br = 80	
6	Rb = 85	Sr = 87	?Yt = 88	Zr = 90	Nb = 94	Mo = 96	? = 100	Ru = 104, Rh = 104, Pd = 106
7	Ag = 108	Cd = 112	In = 113	Sn = 118	Sb = 122	Te = 125	J = 127	
8	Cs = 133	Ba = 137	?Di = 138	?Ce = 140				
9								
10			?Er = 178	?La = 180	Ta = 182	W = 184		Os = 195, Ir = 197, Pt = 198
11	Au = 199	Hg = 200	Tl = 204	Pb = 207	Bi = 208			
12				Th = 231		U = 240		

- In many ways, that's where we are in our modern effort to develop a theory of everything. We know of patterns but don't know the underlying causes. And this, more than anything, is why we have such enthusiasm for the idea of unifying theories—because we've seen many examples of when patterns eventually turned into understanding.

- Our modern understanding of atoms is that they have a particular structure, with a small and concentrated nucleus at the center surrounded by a cloud of electrons. The nucleus is made of protons and neutrons. With protons, neutrons, and electrons, we can build up all of the familiar kinds of matter.

- The situation changed in 1964 and into the early 1970s, when physicists found that the protons and neutrons were made of even smaller particles, which are now called quarks.
- The standard model is one of the 2 biggest steps toward a theory of everything. While nobody claims that we're done, we can regard the standard model as our current best guess of a grand unified theory. Whatever the final theory of everything looks like, the standard model will be part of it.

THE STANDARD MODEL

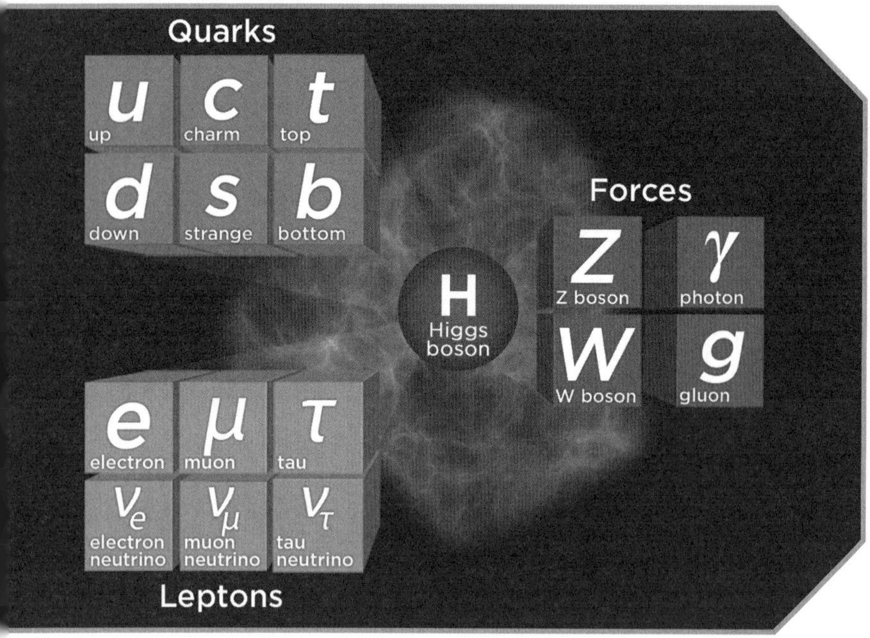

THE KNOWN FORCES

- The 5 known forces are gravity, which keeps us firmly planted on the ground and guides the planets through the heavens; electromagnetism, which covers both electricity and magnetism, as well as light and chemistry; the strong nuclear force, which binds protons and neutrons together in the nucleus of atoms; the weak force, which is responsible for some forms of radioactivity; and the Higgs field, which gives mass to subatomic particles.

- In the late 1960s, physicists showed that the weak force and electromagnetism were really 2 facets of a single thing, much in the same way that electricity and magnetism turned out to be 2 facets of what is now called electromagnetism.

- Therefore, scientists often talk about an electroweak force. So, we might say that the forces are gravity, the electroweak force, the strong force, and the Higgs field. On the other hand, the Higgs field is inextricably tied up with the electroweak force, so maybe it can get tucked under the electroweak umbrella. Under that way of thinking, we have only 3 forces: gravity, the strong force, and the electroweak complex.

- The strong force is the strongest of the known forces. It is incredibly strong over only very short ranges—such as the size of a proton. Once 2 particles are separated by a distance much larger than that, the strong force goes to zero.

- The next strongest force is electromagnetism, which unifies electricity and magnetism into a single force. It's much weaker than the strong force, but it has a different behavior as far as distance is concerned. Two particles experiencing the electromagnetic force will, in principle, feel a force between one another if they are located on opposite sides of the universe. That force will be very small, but it won't be mathematically zero, because electromagnetism has infinite range.

- The next weakest force is the weak force. The natural range of the weak force is about 1/1000 the size of a proton. However, at a separation of a femtometer, it is about 100,000 times weaker than the strong force. At its natural scale, the weak force is actually similar to electromagnetism, and that was the beautiful insight that allowed for electroweak unification.

- Gravity has an infinite range like electromagnetism, but at the femtometer distance scale, gravity is approximately 10^{40} times weaker than the strong force. Gravity is extremely weak—so weak that we have never figured out a way to study it on these very small scales, so it is not covered by the standard model.

- The Higgs field gives mass to particles, so it's not a force in the way that the others are and we can't talk about it in quite the same way. Therefore, we can't compare its strength to the others. The Higgs field turns massless particles into massive ones.

THE STANDARD MODEL

- The quarks are the heavier particles, most commonly found in the nucleus of atoms. There are 6 different types of quarks: up, down, charm, strange, top, and bottom. The up, charm, and top quarks have an electric charge of ⅔ that of the proton. The down, strange, and bottom quarks have an electric charge of –⅓ that of the proton.

- The up and down quarks are found inside protons and neutrons. A proton contains 2 up quarks and a down quark, while a neutron contains 2 down quarks and an up quark.

- All of the quarks experience the electromagnetic force, but they also experience the strong force and the weak force. While the up and down quarks are found in stable atoms, the other 4 quarks are unstable and disappear in much shorter than a second. The only way to study them is to create them in particle accelerators. All 6 types of quarks have been discovered.

- The quarks have very different masses. The up and down quarks have very small masses, less than 1/100 that of a proton. The strange quark's mass is higher than the stable 2, perhaps 20 to 40 times higher. The charm quark is about 30% heavier than the proton, and the bottom quark is more than 4 times heavier than the proton. The top quark is about 185 times heavier than a proton. The top quark was discovered in 1995.

- The leptons are generally lighter than the quarks. Leptons don't feel the strong nuclear force, but they all feel the weak force, and their interaction with the electromagnetic force is mixed.

- There are 2 classes of leptons: charged leptons and neutrinos. The charged leptons, which all carry electric charge, include the electron, the muon, and the tau lepton. The electron weighs about 0.05% that of a proton, while the muon is slightly heavier than 10% of a proton. The tau lepton is just shy of being double the mass of a proton. Like the quarks, the heavier charged leptons decay very quickly, so if you want to study them, you need a particle accelerator.

- Neutrinos don't have electric charge, so they only experience the weak force. Because the weak force is so weak, it means that neutrinos can pass through matter without interacting. Neutrinos are produced in nuclear reactors, and the biggest nuclear reactor is the Sun. The mass of neutrinos is very low. The standard model treats them as having zero mass, but the neutrino masses are actually just incredibly small, not zero.

- The force-carrying particles, or gauge bosons, are responsible for transmitting 3 of the 4 known forces. In the quantum world, forces are caused by the exchange of particles. The gluon is the particle that mediates the strong force, the photon mediates electromagnetism, and the W and Z bosons are responsible for the weak force. Both the photon and gluon are massless, while the W and Z bosons are very heavy—nearly 100 times heavier than the protons.

- There is a hypothetical particle called the graviton that would be responsible for gravity, but it hasn't been discovered.

- The Higgs field and the Higgs boson were proposed in the late 1960s but only discovered in 2012. The Higgs field gives mass to the quarks and leptons, the 2 heavy force-carrying particles, and even the Higgs boson.

- With the standard model, we can explain basically everything we see—from why bread bakes to how stars burn—and we hope one day to unify the electroweak and strong forces into a single force called a grand unified theory. However, the standard model is completely silent on the force due to gravity. For that, we need another theory.

GENERAL THEORY OF RELATIVITY

- In 1915, Albert Einstein revolutionized our vision of the universe. Before his crucial insight, we had to make do with Newton's theory of gravity. Newton unified the gravity of falling objects on Earth with the gravity of stars and planets, but his ideas had stars and planets marching in a stately way through a kind of space that was mathematically flat and unrelated to the objects moving through space.

- Einstein changed all that. He described gravity as being caused by matter and energy actually bending space. Einstein's theory made all kinds of seemingly outlandish predictions. For example, he predicted that colliding stars and black holes would shake the fabric of space and send gravitational waves traveling across the universe. Gravitational waves were indirectly discovered in 1974 and directly observed on Earth in 2016.

- Relativity has been tested to incredible accuracy, but the theory doesn't work when you apply it to the atomic or subatomic world. This means that we still have a way to go in devising a theory of everything. In fact, one of the biggest goals of physics is to somehow first devise a quantum theory of gravity and then merge quantum gravity with the standard model—or, even better, merge quantum gravity with a grand unified theory that unifies all the other known quantum forces. We don't know how to do that, but when we achieve it, we will have taken enormous steps toward a unified theory of everything.

- However, while that unification is a crucial and necessary step, even that is not enough. The standard model has its own holes, and neither theory explains 2 substances called dark matter and dark energy, which are needed to describe the evolution of the universe.

ALBERT EINSTEIN
1879–1955

READINGS

Carroll, *The Particle at the End of the Universe.*
Griffiths, *Introduction to Elementary Particle Physics.*
Kane, *The Particle Garden.*
Lincoln, *Understanding the Universe.*
Veltman, *Facts and Mysteries in Elementary Particle Physics.*

QUESTIONS

1 Why do scientists think that a grand unified theory is possible?

2 What are the key outstanding questions that need to be answered before a grand unified theory might be devised?

2 THE UNION OF ELECTRICITY AND MAGNETISM

We now know that lightning is a form of electricity and that the objects that attract iron are magnets. We also know how the 2 phenomena of electricity and magnetism interact with one another. We have even invented a term that makes this connection obvious: electromagnetism. The development of the formal theory of unified electromagnetism took more than a century. To understand the enormous magnitude of this achievement, we need to immerse ourselves in the science of the 19th century and learn about the details of both electricity and magnetism separately. Only then can we understand how these 2 seemingly different phenomena actually originate from the deeper unified theory that we now call electromagnetism.

ELECTRICITY

- While lightning is the most dramatic example of naturally occurring electricity, our formal understanding of the phenomenon began with a more pedestrian version. This safer version of electricity that first lent itself to formal investigation is static electricity.

- From the turn of the 17th century through the late 1700s, people experimented with static electricity and invented all kinds of gizmos and gadgets to create and concentrate electricity, such as rotating spheres as a way to constantly rub fur on substances like amber. Another useful invention was called the Leyden jar, which was a glass jar with metal foil affixed to the inner and outer surfaces that was used to store electricity.

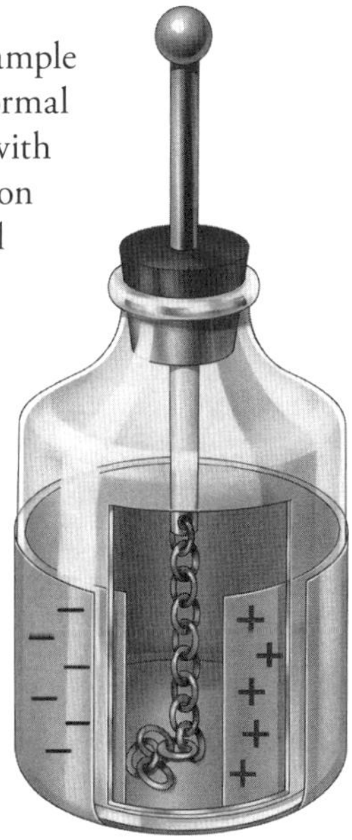

LEYDEN JAR

- Benjamin Franklin famously determined that lightning was an electrical phenomenon. He was the first to divine that there were 2 types of electricity that he called positive and negative. He is also responsible for guessing inconveniently, which is why the electron has a negative charge; it would have been much easier had he given the electron the positive charge, rather than the negative. The electron wasn't discovered until nearly a century after Franklin's death.

- In 1785, our understanding of electricity began to become quantitative, rather than simply qualitative. French scientist Charles-Augustin de Coulomb invented a device that made it possible to carefully measure the force between 2 different charges. Rather than simply knowing that electric charges attract or repel one another, he was able to divine the mathematical formula that governed the strength of the forces.

- He was able to show that the force between 2 objects that carried electrical charge was proportional to the product of the charge the 2 objects carried divided by the distance between them squared. Basically, this says that if we increase the electric charge of either object, we'll increase the force they feel. And as we increase the distance by which they are separated, the force decreases. This is Coulomb's law.

$$F = \frac{1}{4\pi\varepsilon_0}\frac{Q_1 Q_2}{r^2}$$

- In the equation, Q_1 is the charge of the first object, Q_2 is the charge of the second object, r is the distance between them, and ε_0 contains information about how strong the electric force is when you remove considerations of charge and distance. Coulomb's equation only really works for charge concentrated at a small point.

- In 1780, Italian physicist Luigi Galvani discovered that life and electricity were intertwined. In experiments using severed frog legs, he found that if the legs were shocked with electricity, the severed legs would twitch. This showed that life itself was at least partially electrical.

Luigi Galvani
1737–1798

- The subsequent investigation of the role that electricity plays in how life works resulted in a new technology being invented: a battery. The battery simplified how scientists could investigate both electricity and magnetism and greatly accelerated the story and made inevitable the eventual unification that led to electromagnetism.

ALESSANDRO VOLTA
1745–1827

- Alessandro Volta's fertile imagination moved from the idea of animal electricity, which was somehow tied up in the tissues of the frog legs, to the idea of putting dissimilar metals in wine glasses full of brine. This nonintuitive configuration was an example of the first battery. Volta improved his design to a pile of alternating copper and zinc disks, each separated by a brine-soaked disk of cardboard.

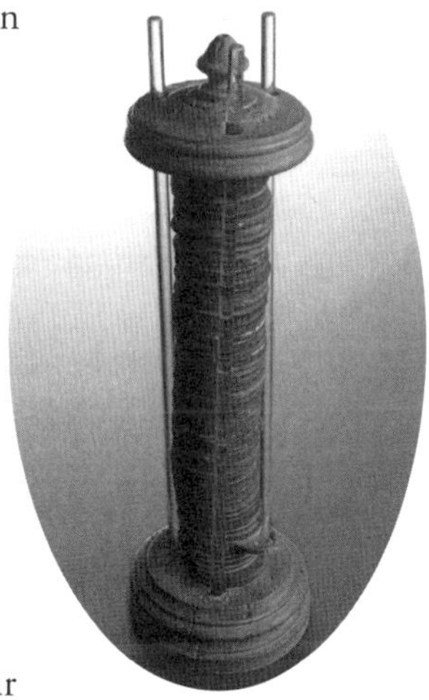

VOLTAIC PILE

- With the invention of the battery, the study of electricity advanced rapidly. Batteries were much more reliable and stable sources of electricity than the earlier electrostatic devices. Batteries could be used to flow electricity through wires for long periods of time. This development was central to our understanding of the deep and fundamental connection between electricity and magnetism.

MAGNETISM

- Humans have known about magnets for a very long time. The earliest writings we have on the subject are those of Thales of Miletus, who lived approximately 2600 years ago. Thales tried to explain static electricity from amber and fur as a form of magnetism. He also experimented with naturally occurring magnets called lodestones, which are naturally magnetized pieces of a mineral called magnetite. Through experimentation, Thales discovered that magnets pick up iron, but not things like gold and silver.

- Magnets became useful tools about 800 years ago, when they were employed first in China, and a few decades later in Europe, as compasses. A magnetized needle could be passed through a cork, which was then floated on water. The needle would point toward the north.

- That leads to a feature of magnetic materials that isn't observed with electrical ones: A magnet has 2 poles. This was first described in a quasi-scientific way in 1269 by French scholar Pierre de Maricourt. He also invented what we now recognize as a compass: a magnetized needle balanced in the middle that spins freely. One end of this needle always points northward.

- From investigations of this nature, we know that each magnet has 2 ends, called a north pole and a south pole. North poles repel other north poles and south poles repel south poles, but north poles attract south poles, and vice versa.

- In 1600, English natural philosopher William Gilbert postulated that the reason a compass needle pointed northward was because the Earth itself was a giant magnet. Gilbert's idea was basically right: The Earth is a magnet whose magnetic field is driven by the motion of magma below the Earth's surface.

- In the early 1800s, after the invention of the battery, the connections between electricity and magnetism began to become clear. In 1819, Danish physicist Hans Christian Oersted discovered that if he took a wire and connected it to a battery, the wire would deflect the needle of a compass. The only way it could do that was if the electric current in the wire generated a magnetic field. This was the first key clue that showed that electricity caused magnetism. The wire makes a magnetic field that has a circular shape, and it doesn't have north and south poles.

Hans Christian Oersted
1777–1851

- This observation took the European scientific community by storm. The next year, in 1820, the observation was shown to the French Academy of Sciences, and André-Marie Ampère began to develop a mathematical theory to describe this. His equation is now called Ampère's law, in which *B* is the symbol for magnetic field, multiplied by a closed path around the wire. That equals the electrical current, *I*, times a constant, μ_0, a magnetic analog, which describes how a magnetic field affects magnets in a medium.

$$\oint \vec{B} \cdot d\vec{l} = \mu_0 I$$

- Ampère's equation has the feature that the left side is all about magnetism and the shape of the magnetic field and the right side is all about electrical current. The 2 sides are joined by an equal sign, which basically means that electricity is magnetism.

- If we know that electricity can make magnetism, can magnetism make electricity? As soon as the world knew of Oersted's work, people set up magnetic fields to see if they induced electricity, to no avail. It took about a decade for the breakthrough to occur.

- In 1831, British scientist Michael Faraday found that while magnetic fields didn't cause electricity, changing magnetic fields did. He found that the electrical voltage generated across a coil of wire was related to a change either in the strength of the magnetic field in the loop or the size of the loop. He wrote down an equation that described this:

Michael Faraday
1791–1867

$$\oint \vec{E} \cdot d\vec{l} = -\frac{d\Phi_B}{dt}$$

- Again, there is an electrical quantity on the left side of the equation and a magnetic one on the right side. Again, electricity and magnetism were related.

- Without this equation, our modern world wouldn't be possible. It explains how both electrical motors and generators work. It also explains how transformers convert the 120 volts that enter your house down to the 5 volts needed to power many electronics. The discovery of Faraday's law was a pivotal moment in the history of technology.

MAXWELL'S LAWS AND ELECTROMAGNETISM

- James Clerk Maxwell was a Scottish physicist who was able to bring together the work of an entire generation of earlier scientists, and in 1861 and 1862, he published what are now called Maxwell's equations.

$$\oint \vec{E} \cdot d\vec{A} = \frac{1}{\varepsilon_0} \int \rho \, dV$$

$$\oint \vec{B} \cdot d\vec{A} = 0$$

$$\oint \vec{E} \cdot d\vec{S} = -\frac{d}{dt} \int \vec{B} \cdot d\vec{A}$$

$$\oint \vec{B} \cdot d\vec{l} = {}_0 \int \vec{J} \cdot d\vec{A} + {}_0\varepsilon_0 \frac{\partial}{\partial t} \int \vec{E} \cdot d\vec{A}$$

- These 4 equations incorporate our understanding of both electricity and magnetism. These equations have some calculus in them, but to understand the most important points, keep in mind that E stands for electric fields and B stands for magnetic fields. J stands for electric current and ρ stands for distributions of electric charge. The S, A, and V are just geometry.

- The first equation is called Gauss's law for electricity, and the second is Gauss's law for magnetism. Gauss's law is an extended form of Coulomb's law. You don't see any mixing between electricity and magnetism in these 2 equations.

- However, the third equation is Faraday's law, and the fourth one is Ampère's law. Note that both of these equations mix electricity and magnetism. Taken together, these 4 equations show that electricity and magnetism are really the same.

- You can use Maxwell's equations to come to a crucial insight. These equations can be used to derive a classical wave equation that shows that both electricity and magnetism are waves. The equations can also be used to derive an equation that shows that light is related to both electricity and magnetism and that light is a wave. In short, Maxwell proved that light is an electromagnetic wave.

- An electromagnetic wave oscillates just like any other wave. Furthermore, the electric and magnetic components both travel at the same speed, which is the speed of light. From Maxwell's equations, we can derive a wave of electricity and a wave of magnetism.

- In our efforts to find a unified theory of the universe, this is an incredible achievement. We can show that electricity and magnetism are the same and that light is electromagnetic. We can also use electromagnetism as a basis of chemistry.

READINGS

Forbes and Mahon, *Faraday, Maxwell and the Electromagnetic Field.*
Gibilisco, *Electricity Demystified.*
Griffiths, *Introduction to Electromagnetism.*
Mahon, *The Man Who Changed Everything.*

QUESTIONS

1 How are electricity and magnetism the same, and how are they different?

2 Why do Maxwell's equations prove that electricity and magnetism should be replaced by a single theory called electromagnetism?

3 PARTICLES AND WAVES: THE QUANTUM WORLD

One of the oldest arguments in physics is on the nature of light: Is light a particle or a wave? Waves have wavelengths, frequencies, and amplitudes and extend over a large volume; particles have a well-defined location, without the trappings of a wave. Maxwell confirmed that light is a wave, but a fully correct answer is much more complicated. If a person proposes one or the other theory, at a minimum, the proposal needs to explain some of the things we see light do, such as reflect and refract (the effect whereby light bends when it passes from air to a transparent solid).

LIGHT: PARTICLE OR WAVE?

- While early discussions about the nature of light began in Greek antiquity, the arguments were essentially philosophical until around the time of Newton, who was a proponent of the particle view. If light is a particle, it is easy to explain how light reflects. If you assume that light travels faster through a solid than air, the particle idea can explain the observed bending of light as well.

- We now know that this assumption is backward: Light travels more slowly through solid matter than it does air. But that wasn't known at the time. Newton's extraordinary scientific reputation led the particle theory of light to be favored for a very long time.

- However, the wave idea had its adherents. English scientist Robert Hooke and Dutch mathematician Christiaan Huygens—both roughly contemporaries of Newton—worked out a wave theory of light. This theory can also explain reflection and refraction, although the refraction explanation required that light move more slowly in solid matter rather than gaseous matter.

- We know now that this is correct; however, for about 150 years, there was no measurement to prove it. Huygens postulated that there was a substance called the luminiferous aether that allowed light to travel as a wave. While we no longer believe in the aether, theorists continued to talk about it into the late 19th century. The lack of any evidence for aether was an objection to the wave theory.

- The argument persisted until about 1800, when a series of experiments by English polymath Thomas Young demonstrated rather persuasively that light was actually a wave. One of the properties of waves that is very different than particles is that they can interfere with one another.

- Young's experiment used an opaque barrier with 2 adjacent slits. When light illuminated a single slit, it passed through the slit and spread out. That's true of the other slit as well. If light is a wave, you'd expect it to look like the ripples in a pond when a stone is dropped in the water.

- On the other hand, both slits generate these ripples, which means that there would be 2 sources of waves. And waves can interfere with one another. There would be spots that are peaks of one set of waves and troughs of the other, which would cancel out. Cancelling out light means that there would be dark spots on a distant screen. Similarly, there would be places where the peaks of both sets of waves would appear at the same location, resulting in bright spots on the distant screen.

THOMAS YOUNG
1773–1829

- And that's exactly what we see: bright spots and dark spots. Young's double-slit experiment is the definitive and inarguable demonstration that light is a wave. Maxwell's equations of the 1860s and 1870s simply solidified the situation.

- That was the status quo for a while. However, in 1887, German physicist Heinrich Hertz did an experiment that was puzzling. He attached 2 electrodes to both ends of a very strong battery. If he did so, occasionally a spark would jump between the electrodes. He noticed that if he illuminated the electrodes with ultraviolet light, the sparks occurred more easily than if the same experiment was done in the dark.

- The initial explanation was that the electromagnetic field was knocking out some of the electrons from the electrodes, which then started a chain reaction. Actually, even now that's what we think. But there's more.

- People started playing around with the parameters that define the light: the color of the light (wavelength) and the brightness (amplitude). They saw that the color of light seemed to matter. For all materials, certain colors, visible or not, would induce a spark, while other colors wouldn't.

- The colors of the rainbow are red, orange, yellow, green, blue, indigo, and violet. The color of light is determined by the wavelength, with red having a long wavelength and violet having a shorter one. The wavelength and frequency of light are related by the mathematical relationship that frequency times wavelength equals the speed of light. That means that red light is low frequency, while purple light is high frequency. This says that high-frequency, short-wavelength, light can cause a spark, while low-frequency and long-wavelength light can't.

- Another observation is that if you shine low-frequency light on the electrode, you can crank the brightness higher and higher and you get no spark. This is where things get tricky. Young's double-slit experiment established that light was a wave, and waves have certain properties, one of which is that the energy of the wave is dependent on the amplitude. Thus, if you cranked up the brightness, you should be hitting the electrode with more energy—if light were a wave, anyway.

- The early part of the 20th century was a confusing time for scientists. But the situation was hugely clarified in 1905, when Albert Einstein published a paper that explained why some wavelengths cause electrodes to spark while others don't. It is called the photoelectric effect.

- Einstein solved the problem by imagining that light was actually a particle that retained a little bit of its wave nature. Einstein called these particles light quanta, but now they are called photons.

- Einstein began by observing that the energy of the photons is proportional to their frequency. Specifically, the relationship is $E = hf$, where E stands for energy, f stands for frequency, and h is Planck's constant. This idea of relating energy and the frequency of light didn't originate with Einstein; in fact, Max Planck had proposed a variation of the idea in 1900 to explain some mysteries that arose when people studied the color of objects heated until they glowed.

THE ELECTROMAGNETIC SPECTRUM

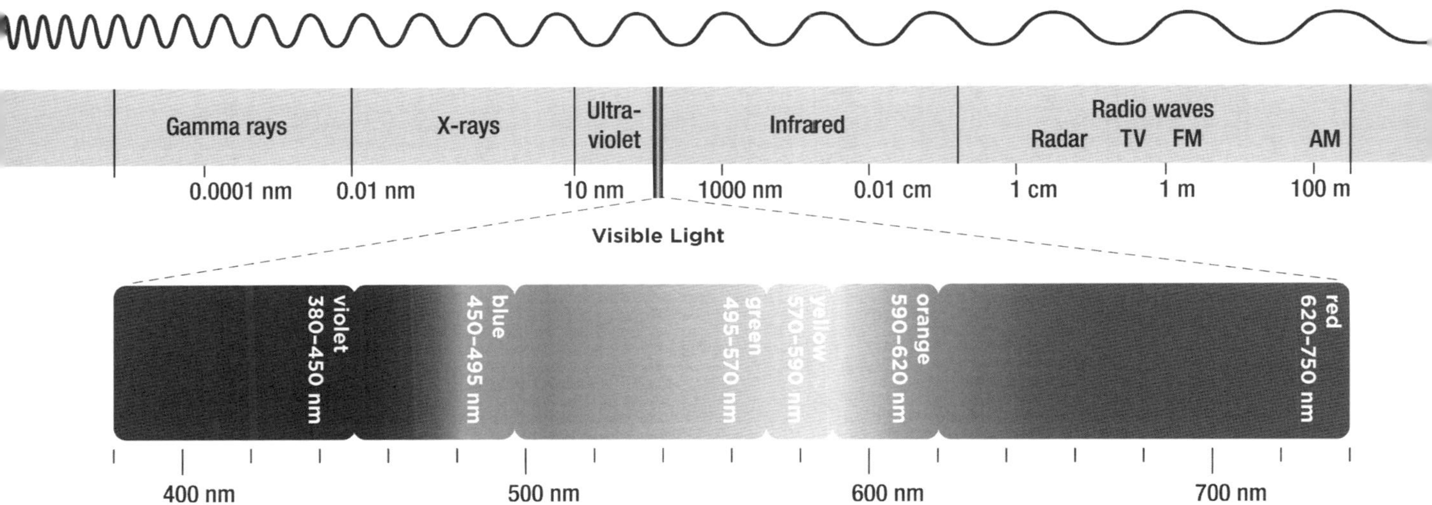

- But Einstein used the idea in a novel way, and it explained the observed relationship between sparks and the color of light that hit the electrodes. The basic idea is pretty simple. Remember that the electrodes were made of atoms and that atoms are surrounded by electrons.

- Think of atoms like a solar system, with a nuclear sun and planetary electrons. The electric field of the nucleus holds onto the electrons with a certain force, or certain amount of energy. It requires a certain amount of energy to pull an electron off the atom.

- And that was the key insight. If Einstein was right, and the energy of a photon was proportional to the frequency, then only light with enough energy could knock electrons off the electrode. That meant that only high-frequency light could cause a spark, and the whole thing was explained. And all this showed that light had to be a particle, because if it were a wave, increased brightness would have caused a spark, but if light came in distinct quantized particles with distinct energy, each photon could knock out an electron and initiate a spark.

- So, we were left with a mind-bending conundrum. Young's double-slit experiment had proved that light acted like a wave. The photoelectric effect proved that the photon was a particle. And because a wave is nothing like a particle and a particle is nothing like a wave, that's a huge problem. It's as if the photon were both a wave and a particle.

- There's a way to do the double-slit experiment but with the brightness of the light turned down so much that if light were a photon, only one photon would be emitted at a time.

- If photons are waves, then you'd expect that each photon goes through both slits and that the waves will interfere in the way that Young found and hit the distant screen. Each photon would give just a very dim contribution, but as you added up more and more photons, you'd eventually end up with the same interference pattern that Young found more than 200 years ago. And if the photon acted like a particle, you'd see that the photon would appear in one spot at the distant screen.

- While photons hit the screen one at a time, they don't hit everywhere on the screen. There are places where lots of photons hit the screen and places where none hit the screen. And after we send tons of photons through the system, we see Young's double-slit distribution: The photons end up looking like a wave pattern.

- Wave physics governs how photons travel, but they are detected as particles. The best-regarded explanation for this is that the motion of photons is governed by waves—in fact, the waves are thought to describe probabilities—and when we detect the particle, the wave collapses instantaneously.

THE QUANTUM THEORY OF MATTER

- At the turn of the 20th century, English physicist J. J. Thomson had discovered the electron and thought that he understood what an atom looked like. He imagined that electrons were small and compact objects with negative electrical charge and that they were embedded in a massless and positively electrically charged goo. His model was called the plum pudding model.

- Thomson's vision of the atom was considered to be a reasonable idea for a little more than a decade, until 1911, when New Zealand physicist Ernest Rutherford shot a form of radiation called alpha particles at a foil of gold and saw that some of the alpha particles bounced back.

Ernest Rutherford
1871–1937

THOMSON'S MODEL

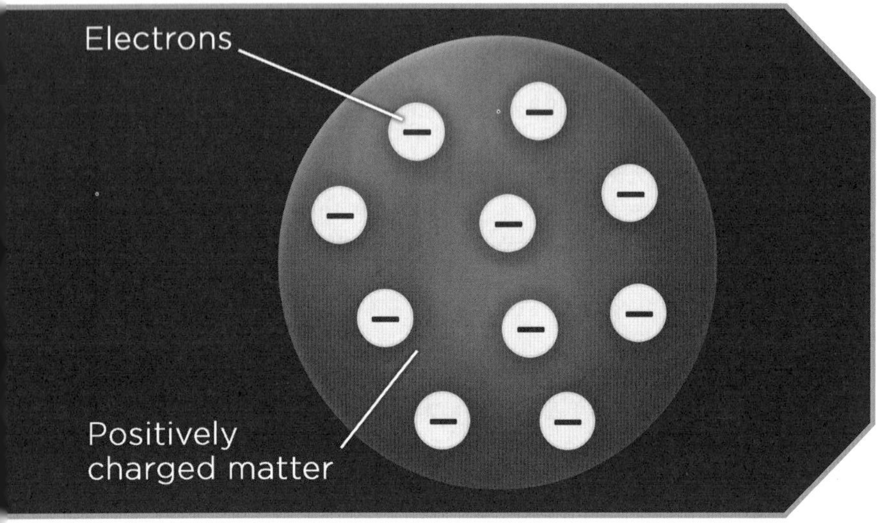

RUTHERFORD'S MODEL

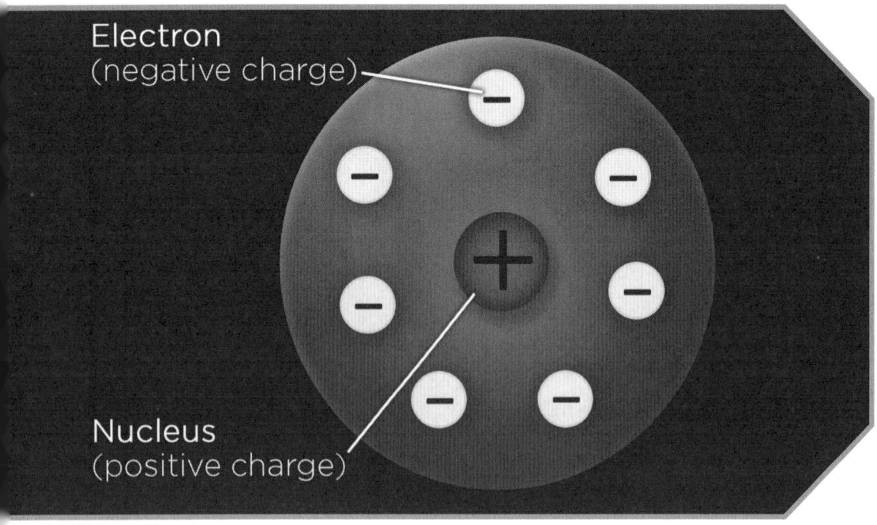

- It's difficult to imagine Thomson's goo reflecting anything, so the plum pudding idea was ruled out. Rutherford used his data and came up with something approaching a modern model of the atom, with all the positive charge gathered together at the center of the atom and with the electrons swirling around that nuclear core—called a nucleus—at relatively large distances. Rutherford's model of the atom looked very much like a solar system, with a nuclear sun and planetary electrons.

- One big problem is that the theory of classical electromagnetism said that Rutherford's idea was impossible. Classical electromagnetism explains in very specific detail what happens when an electric charge is accelerated. And if an electron is orbiting a nucleus, it is constantly being accelerated toward the nucleus.

- Classical electromagnetism predicts that an accelerated charge will emit electromagnetic radiation—basically, it gives off electromagnetic energy. And if the electron is giving off energy, then that means it is losing energy. And if it loses energy, then it will spiral down into the nucleus of the atom. Because we know that atoms don't spiral down into the nucleus, that is a problem.

- Another problem with the Rutherford atom was that Rutherford didn't say much about the electrons swirling around the nucleus. Presumably, they could orbit near the nucleus or far away or anywhere in between. And when the electrons moved from one orbit to another, they had to change their energy and would therefore emit electromagnetic radiation. It turns out that for many atoms, the electromagnetic energy they would emit would be in the form of visible light.

- Given that there were no constraints on the orbits of the electrons, that means that when they changed from one orbit to another, there were no constraints on the wavelength of light they would emit. That means that when we look at a glowing gas, all colors should be present. However, that's not what we see when we look at a glowing gas. If we take that light and run it through a prism, we see that only certain colors exist.

- About 2 years later, in 1913 Danish physicist Niels Bohr embraced the Rutherford model and then made an additional assumption. He simply said that the electrons flying around the nucleus of an atom orbited in a series of fixed orbits, which he labeled 1, 2, 3, etc.

- According to this hypothesis, the electrons could jump between orbits but never exist in between them. Each orbit would have a different (and fixed) energy, and that means that when an electron changed from one orbit to another, the energy of the photon emitted would come in discrete chunks.

- Another important feature is that because all electrons had to exist in an orbital, once electrons moved to the first orbital, they could not lose any more energy, because the first orbital had the lowest energy. This feature protected the atom against electrons radiating energy and spiraling down into the nucleus.

BOHR'S MODEL

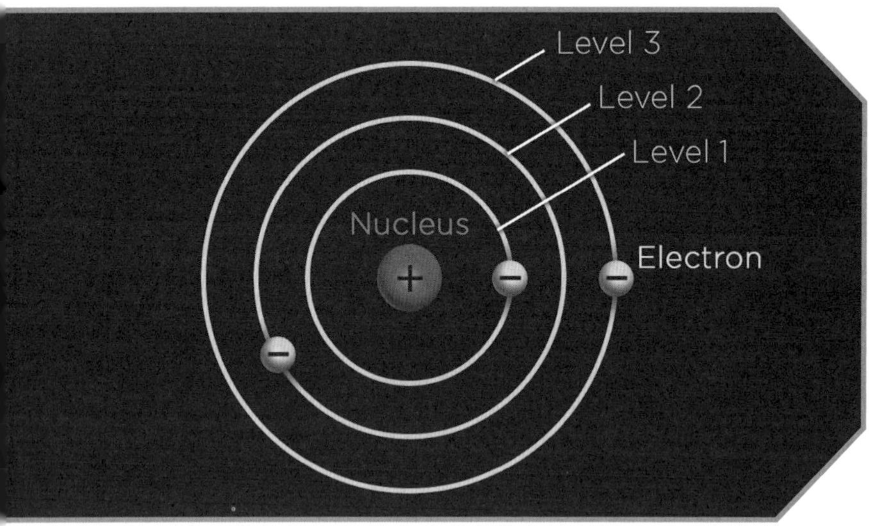

- This idea that there were fixed orbits also explained why hot gasses emitted light at fixed wavelengths. Because electrons could only transition between a set of discrete orbits, that means that there were only specific energies that photons could have. At least qualitatively speaking, that explains the spectrum issue. Bohr was actually able to predict the spectrum of hydrogen essentially perfectly. But there was no reason to suppose he was right; his model needed a motivating explanation.

- In 1924, French physicist Louis de Broglie wondered whether the electron had a wave nature, too. If the electron was like a photon and was both a particle and a wave, then the way to prove that was to demonstrate that electrons could interfere with one another. This was accomplished by scattering electrons off crystals.

- The wave nature of the electron explained the Bohr atom. Bohr's model of the atom had a series of energy levels that were discrete, but it didn't explain why, while de Broglie's idea did. If an electron is a wave and it has to fit in an orbit surrounding an atom, the wave must then "fit" in the orbit. That means that the orbit must be a fixed number of wavelengths. The first orbital would fit 1 wave, the second orbit would fit 2, the third 3, and so on. The subatomic world did seem to require a blend of particles and waves.

- To describe the behavior of the electron as a wave, Austrian physicist Erwin Schrödinger invented in 1925 what is now called the Schrödinger equation, in which the psi (ψ) term describes the particle and its kinematic properties, the H with the hat over it is a way of extracting the energy of the particle, and the term on the right is the partial derivative with respect to time of the psi function (which, if the situation is static, is also the energy). This compact equation is the explanation for all of atomic physics and all of chemistry.

$$\hat{H}\left|\psi(t)\right\rangle = i\hbar\frac{\partial}{\partial t}\left|\psi(t)\right\rangle$$

THE WAVE FUNCTION

- Psi embodies all of our understanding of the electron in the atomic realm. It is a wave function. The wave function describes the electron—both its location and its energy. If you solve the Schrödinger equation for a free particle, which just means a particle in empty space with no nearby electric fields, the wave equation is a wave, wiggling up and down.

- However, in the vicinity of an atomic nucleus, the wave function takes on different shapes. Stealing some language from the Bohr atom, the 2 lowest-energy orbitals are spherical, while the third is shaped like a dumbbell. Higher-energy orbitals take on increasingly complicated shapes.

- But the shapes aren't as important as the physical significance of the wave function. For example, what does the dumbbell-shaped one mean? This is a question that still puzzles scientists. It could be that it describes where an electron is, but that doesn't seem to be quite right.

- Over the period of 1925–1927, Niels Bohr and Werner Heisenberg devised an answer, now called the Copenhagen interpretation. They proposed that the wave function was related to the probability of finding an electron at a specific location. Technically, they found that if they squared the wave function, the result was the probability.

- But probability doesn't work the same in the quantum world. It seems as if the electron is simultaneously everywhere that the wave function says it is; however, when you detect it, the location of the electron is found in a certain location, with the probability determined by the square of the wave function.

- This is reminiscent of Young's double-slit experiment of photons, where the photon seemed to act like a wave when it was traveling and not being looked at, but instantly turned into a particle with a specific location when it was detected.

READINGS

Gamow, *Thirty Years That Shook Physics.*
———, *Mr. Tompkins in Paperback*, chap. 7–10.
Gilmore, *Alice in Quantumland.*
Gribbin, *In Search of Schrodinger's Cat.*
Griffiths, *Introduction to Quantum Mechanics.*
Susskind and Friedman, *Quantum Mechanics.*

QUESTIONS

1 In what way is the quantum realm different than the macroscopic world? What are the key features that make quantum behaviors appear in the microcosm that elude us in the macroworld?

2 Does Schrödinger's equation apply only to atoms, or does it apply to other situations?

4 EINSTEIN UNIFIES SPACE, TIME, AND LIGHT

The theory of special relativity offers all kinds of weird predictions, such as people experiencing time and space differently depending on their relative speed. However, once you embrace the idea that the distinction between space and time is illusory and space-time is what matters, you've gotten a lot of the way toward understanding what special relativity is all about. Then, if you add the idea that all objects travel at only one speed in space-time, you have everything you need to understand relativity at an intuitive level.

VELOCITY

- There are some facts that transcend the question of whether we are thinking about the classical or the quantum world. For example, no matter whether you are measuring the speed of a classical electromagnetic wave or the speed of a photon, you get an identical number—specifically 186,282.39 miles per second. That's a huge number, enough to circle the globe about 7.5 times in a single second.

- In addition, the speed of light is known very precisely, to 9 digits when expressed in meters. Rounded up, it's 300 million meters per second.

- Speed isn't absolute. Different people will disagree on an object's speed.

- If we were able to use Maxwell's equations to predict the speed of light, then the question is, the speed with respect to what? And trying to answer that question changed everything.

- There were many ideas on this in the late 1800s. However, the most common idea was one first championed by Christiaan Huygens in the late 1600s, one of the earliest proponents of the wave theory of light. This was the idea confirmed by Young's double-slit experiment, performed in 1800, which conclusively demonstrated light's wave nature.

- Huygens's idea was that the universe was filled with a substance called the luminiferous aether. Basically, in the same way that water transmits water waves and air transmits sound waves, aether was supposed to transmit light waves. And the speed of light would always be measured with respect to the aether. So, if that's true, as long as you could move with respect to the aether at the same speed that light does, then, at least in principle, you could see light stand still.

- This idea was state of the art until 1886, when American physicists Albert Michelson and Edward Morley performed an experiment that showed that the aether didn't exist. But if no aether existed, what was it that allowed light to travel? And what was the speed of light determined with respect to?

SPECIAL RELATIVITY

- Albert Einstein imagined what would happen if 2 people were in outer space and were moving toward each other. How would we know if the first person was stationary and the second person was moving toward the first? Or if the second person who was stationary and the first was moving toward the second? Or maybe that both people were moving off toward Pluto, but one was moving faster, so they kept getting closer? He decided that there was no way, even in principle, that we could know which of these scenarios was true.

- This led him in 1905 to 2 postulates from which he built a new theory of motion. His postulates were:
  1. The laws of physics are the same for everyone. The universe plays no favorites and doesn't decide on who is moving and who isn't.

  2. The speed of light is measured to be the same for everyone. Two people who are moving at very different speeds will both measure the speed of light to be the same.

- Although the second postulate is unconventional, scientists have done many tests of the universal speed of light in the century since Einstein, and there is no doubt that it is true to very good precision. And even if you can't test that hypothesis directly, you see what consequences it predicts and then check them. That turns out to be much easier.

- The theory derived by Einstein is called the special theory of relativity because it addresses the special case when everything is moving at constant speed.

- Suppose that you have a train car with mirrors on the walls. Also suppose that the width of the train is just given by the symbol w. If you're just sitting in the train and you wanted to know how long it takes for a beam of light to cross the train, it's just the width divided by the velocity of light, c. If we call the time it takes to be t with a subscript s (meaning stationary), then the time is $t_s = w/c$.

- Now suppose that you're outside the train and you see that the train is rolling along. That means that the mirrors are moving while the beam of light is moving from one to the other. While it is moving, the train travels some distance *d* along the track. The light will travel along the hypotenuse of a right triangle, with one leg being distance *w* and the other distance being *d*.

- Using the Pythagorean theorem, the hypotenuse, *h*, will be $h = \sqrt{w^2 + d^2}$. If light is moving at the speed of light, then the time it takes to go from one mirror to the other will be $t_m = h/c$, in which t_m is the time for the person who sees the train moving. After doing some substitutions and other math, we get the following:

$$t_m = \frac{t_s}{\sqrt{1 - \left(\frac{v}{c}\right)^2}}$$

- Our intuition is that the moving time and stationary time should be the same, but we see that this isn't true. The whole square root term is a core term in special relativity called gamma (γ).

$$\gamma = \frac{1}{\sqrt{1 - \left(\frac{v}{c}\right)^2}}$$

$$t_m = \gamma t_s$$

- If we enter into the equation a velocity that is equal to the speed of light, we find that gamma is undefined. And that is at least the mathematical reason why we say that nothing can go faster than the speed of light: If we put in a velocity that is greater than the speed of light, we get an imaginary number.

TIME AND SPACE

- The time equation we just derived basically says that a stationary observer experiences a shorter amount of time than a moving observer. That is a mind-blowing and counterintuitive idea. But it's been tested and is correct.

- GPS satellites would not work accurately enough if we did not adjust for how their high speed across the sky causes their clocks to run slower.

- An observer who sees an object moving experiences a longer amount of time than a person who sees the object as stationary.

- Classical physics makes one prediction and special relativity makes a very different prediction, and special relativity is right.

- Time isn't the only thing that is different for moving and stationary objects. While the time experienced by a person seeing a moving thing is longer than for a person seeing a stationary thing, the physical length of that moving thing is shorter, but only in the direction of motion.

- It's more difficult to measure this effect directly, but it turns out that if you combine Einstein's theory of relativity with Maxwell's equations, the unification of electricity and magnetism gets much deeper. In Maxwell's equations, the 2 ideas are inextricably tied together but still distinct. When you add relativity, the distinction evaporates, with moving and stationary observers disagreeing about the amount of electricity and magnetism and the 2 phenomena become truly indistinguishable.

MASS AND ENERGY

- Perhaps the most famous equation in physics is Einstein's equation $E = mc^2$, in which E stands for energy, m stands for mass, and c is the speed of light. In words, it says that energy equals mass times a constant—specifically, the speed of light squared.

- A constant is basically just a conversion factor, so the equation essentially says that energy and mass are the same thing. You can convert one into the other and back again. But how much mass equals how much energy? It turns out that a little mass is a huge amount of energy; mass is ultra-concentrated energy.

- People misuse the $E = mc^2$ equation all the time. That's because it is a special case of a more general equation. It is for a stationary object. The more general equation is this:

$$E^2 = (pc)^2 + (mc^2)^2$$

- The variables are the same except now we have p, which stands for momentum, which is kind of a motion energy. If there is no motion energy, then $p = 0$, and if you do that, you get back the $E = mc^2$ equation. And if you have a particle with zero mass, such as the photon, you can insert $m = 0$ and get that $E = pc$, which is the right equation for a photon. For an object with some mass and some momentum, you have to use the full equation; otherwise, you get wrong answers.

- Another misconception is that a particle's mass gets bigger as speed increases. But that's not true. It's a way to give an intuitive feel for one of the weird predictions of relativity theory: that you can't go faster than the speed of light, which is true.

- That's not what you might have predicted using your intuition that you've developed by going at pretty slow speeds. After all, at familiar speeds, you can always go faster. However, as you get close to the speed of light, it becomes harder and harder to increase your speed. An object acts like it is getting more massive because, in our experience, it's harder to push a heavier object than a lighter one.

- Scientists even invented a term for this: relativistic mass, which we distinguish from rest mass, which is the mass of an object that isn't moving. Relativistic mass gives people wrong intuitions, but it exists because the idea that there is a maximum speed is foreign to us.

- It's not mass that is increasing—it's inertia. And at low speeds, mass and inertia are basically the same thing. Relativistic mass combines our low-velocity intuition of the interconnections of mass and inertia with the fact that you can't go faster than light. Thus, relativistic mass is a way to modify people's intuition. The downside is that it comes with misconceptions, and people who pursue higher education in physics have to unlearn this pedagogically useful and yet wrong concept.

SPACE-TIME

- If you have an incident that occurs at a particular location and time, relativity tells you what location and time will be seen by a person who is moving with respect to you. Your mix of space and time will differ from your moving friend's.

- In 1905, Einstein first wrote down his equations that transformed the space and time coordinates from one observer to another, but 2 years later, German physicist Hermann Minkowski figured out the equation's deeper meaning.

HERMANN MINKOWSKI
1864–1909

- Minkowski realized that Einstein had unified space and time into what is now called space-time. Essentially, space and time are now known to be the same thing. Furthermore, he realized that Einstein's transformation equations were just rotations in space-time.

- A person who is stationary is space only experiences translations through time and none through space. On the other hand, a person moving faster and faster compared to you experiences times that are shorter and shorter. And when you get to the speed of light, the time experienced by that person eventually goes to zero. So, at the speed of light, there is only movement through space and no movement through time. The only changes to experience are in space, while there are no changes in time. And if a person is moving at a velocity through space that is less than the speed of light, that person is experiencing both changes in space and time.

- A person's space-time vector—that is, the combined experience of both space and time—is unchanged. One person might experience more space and less time, while another might experience more time and less space. But the combination—the length of the space-time vector—is something everyone agrees on.

- This means that a person's velocity through space-time is constant and the person's speed is the speed of light. A stationary person is moving through time at the speed of light, while a person moving 100% through space and not at all through time is moving at the speed of light through space.

- That is a deep, fundamental, and satisfying answer to the question of why you can't go faster than the speed of light. It's because everything is always going at that one single speed. And it also shows how space and time are one and the same and that how you experience the 2 is just a matter of perspective.

READINGS

Bennett, *What Is Relativity?*
Einstein, *Relativity*, part I.
Gamow, *Mr. Tompkins in Paperback*, chap. 1–2.
Gardner, *Relativity Simply Explained.*
Styer, *Relativity for the Questioning Mind.*
Wolfson, *Simply Einstein.*

QUESTIONS

1 What differences between space and time are the most difficult to reconcile when people realize that they are just components of a single idea called space-time?

2 What makes it impossible to go faster than the speed of light?

RELATIVISTIC QUANTUM FIELDS AND FEYNMAN

5

This lecture starts with the successful blending of relativity and quantum mechanics into relativistic quantum mechanics, and that leads into the world of quantum field theory, which underlies all modern particle physics theories. You will learn about Feynman diagrams, which are scientific hieroglyphs that teach us a lot about what is going on. You will also learn about perturbation theory, which is a clever way to solve an otherwise-intractable problem. The larger unification involves replacing a stand-alone classical theory of electromagnetism, which involved a continuous field, and instead combining electromagnetism with a quantum formation to create a synthesis known as quantum electrodynamics.

RELATIVISTIC QUANTUM MECHANICS

- The previous 2 lectures addressed impressive insights into both the quantum world and the realm of the ultrafast. However, these true theories were separate, when it was obvious that they needed to be merged. After all, small things can go very fast, and in that situation, both theories must apply.

- The first big step forward occurred in 1928, when British physicist Paul Dirac created what is now called the Dirac equation. This equation successfully blended quantum mechanics and special relativity, but it also incorporated the quantum mechanical spin of the particles involved.

$$\left(\beta mc^2 + c\left(\sum_{n=1}^{3} a_n b_n\right)\right)\psi(x,t) = i\hbar \frac{\partial \psi(x,t)}{\partial t}$$

- Particles such as protons and electrons are both electrically charged and act very much as if they are spinning. A spinning charge will create a magnetic field, so Dirac's equation explicitly brought magnetism into the picture. Perhaps even more interesting, quantum mechanical spin was necessary to build a coherent theory.

- Dirac's theory also predicted the existence of antimatter. In essence, when he had finished his mathematical wizardry, he ended up with a solution that looked like $(\text{equation})^2 = 1$. To solve that, you take the square root of both sides. But that gives you an answer like (equation) = ±1, which means +1 and –1 are both answers.

- Dirac insisted that his equation was so beautiful that all solutions needed to be taken seriously. He thought that it was clear that the numerically positive solution described the ordinary matter we see in the universe around us. In the specific problem he was studying, the numerically positive solution would represent the electron, which has a negative charge. He wasn't sure exactly what the numerically negative solution represented, but one possibility, he thought, might be the proton, which has a positive charge.

- We now know that the numerically negative solution wasn't the proton. It was a positively charged particle, but with the mass and other characteristics of the electron. We now call this opposite-charge version of the electron a positron, which is an example of what we call antimatter. The positron was discovered in 1932 by American physicist Carl Anderson.

- About 80 or more years after the antielectron was discovered, we know that all forms of matter have antimatter analogs. In addition to the antielectron, there are also antiprotons, antineutrons, and a bunch of other antiparticles whose matter analogs weren't even known in the 1930s. Quarks and leptons were identified later, and it turns out that there are also antiquarks and antileptons.

- Antimatter is essentially identical to regular matter in many ways. For example, a positron has the same mass as an electron, while the antiproton has the same mass as a proton. The difference between matter and antimatter is that they have opposite electric charge. Thus, while the proton has a positive charge, the antiproton has a negative charge.

- Another interesting consequence of Dirac's theory is that matter and antimatter are antagonistic substances. If you combine matter and antimatter, they annihilate in a huge burst of energy.

- But perhaps the most important outcome of Dirac's equation is that it actually heralded a second quantum revolution. In the first quantum revolution, embodied by the Schrödinger equation, de Broglie's hypothesis, and Einstein's explanation of the photoelectric effect, matter was shown to come in quantum chunks. Electrons flit around the nucleus of atoms in fixed and quantized orbits. Photons are quantum messengers of electromagnetism.

- But one thing wasn't quantized: the electric field of the nucleus. In Schrödinger's equation, the nucleus's electric field was entirely classical. Maxwell might not have understood quantum mechanics, but he would have recognized how early quantum mechanics handled the electric field of the nucleus: The nucleus was still a classical, continuous field.

- Dirac and others who further developed his theory expressed the electric field caused by the nucleus as a huge buzzing swarm of photons. In this view, if you try to scatter an electron off an atom, rather than encountering a smooth and continuous electric field, the electron would dive into a maelstrom of individual and distinct photons around the nucleus of the atom. Some photons would interact with the electron and divert its path.

- Adding up the effects of this complicated set of possibilities is pretty difficult, and we need to invoke a mathematical trick. This trick hinges on the fact that not all photons are created equally. Some photons would transfer a lot of momentum, while other photons would transfer only a modest amount. The photons with lots of momentum will have the biggest impact on the electron's trajectory.

PERTURBATION THEORY

- It turns out that Dirac's equation is very difficult to solve. In fact, it is currently insoluble in any exact way. But this isn't such an unusual phenomenon. Many equations can't be solved exactly. Scientists and mathematicians have invented powerful techniques to address this problem.

- One very effective technique is to take a difficult problem and replace it by an easier one that is very similar. As long as the differences between the original problem and the approximation are small enough, you can make reasonably accurate predictions.

- Perturbation theory is used to simplify solutions in quantum field theory. Precise solutions to the mathematics of quantum field theory are not even possible, so we need working approximations that we can make as precise as needed.

QUANTUM ELECTRODYNAMICS

- Quantum electrodynamics (QED) is the name of the modern theory that governs the interaction between electrically charged particles. Dirac's theory has most of the core components—specifically quantized bits of charge-exchanging quantized photons—but the modern theory is a little different, mostly in mathematically technical ways that aren't necessary to understanding the crucial points.

- Imagine what happens when we scatter 2 electrons off one another. Classically, the electric field from each electron repels every other electron. The net result is basically that they bounce off one another without touching. However, that way of thinking about electron interactions is outdated.

- In QED, there isn't a smooth and continuous electric field. Instead, the 2 electrons are constantly emitting and absorbing a bath of photons. As the electrons approach one another, those photons can jump from one electron to another.

- In QED, as 2 particles approach one another, they exchange photons. In the simplest case, the simplest approximation, one charged particle shoots a photon at the other particle and both charged particles recoil. That's called leading order.

- But that's not the only thing that can happen. At next to leading order, which is the next term in the perturbative expansion, sometimes when the 2 particles collide, they exchange 2 photons.

- Sometimes a particle emits a photon, which it reabsorbs after scattering. Or maybe one of the particles emits and then reabsorbs it before it scatters off the photon from the other particle.

- This can happen in many ways, with both particles participating and in both the incoming and outgoing pieces. In a sense, anything can happen, all at the same time, so it's a question of how many layers of approximation you want to include.

- The possibilities become complicated. Sometimes photons can temporarily convert into matter/antimatter charged particles before recoalescing back into a photon. And we can get much more complicated scenarios, with charged particles emitting and absorbing particles before and after the collision. Some of those exchanged photons can make charged particle pairs that interact with other photons.

- One of the tenets of quantum electrodynamics is that you have to average all possible configurations of different ways 2 particles can interact. In fact, one of the weird aspects of quantum mechanics is that you can't say which specific interaction happened. They all did, and they all need to be taken into account.

- When you add them all up, you find that the sum of all interactions looks very much like the classical picture. In this way, the simplified quantum idea of individual photon exchanges can connect to the familiar classical world of a smooth electromagnetic field.

PERTURBATION THEORY AND QED

- Each of the many different ways the electrons can interact can be represented by a doodle called a Feynman diagram, which help us understand us what is going on.

- When we are talking about 2 electrons bouncing off one another, there are many doodles to consider, and to do a proper and 100% accurate calculation, you need to add up the contributions from every single one of these doodles. And that is a pain in the neck. In fact, nobody knows how to do it. So, we need a better way.

- Luckily, there is a mathematical trick that can help us out—by separating the various doodles into different categories of importance.

FEYNMAN DIAGRAM

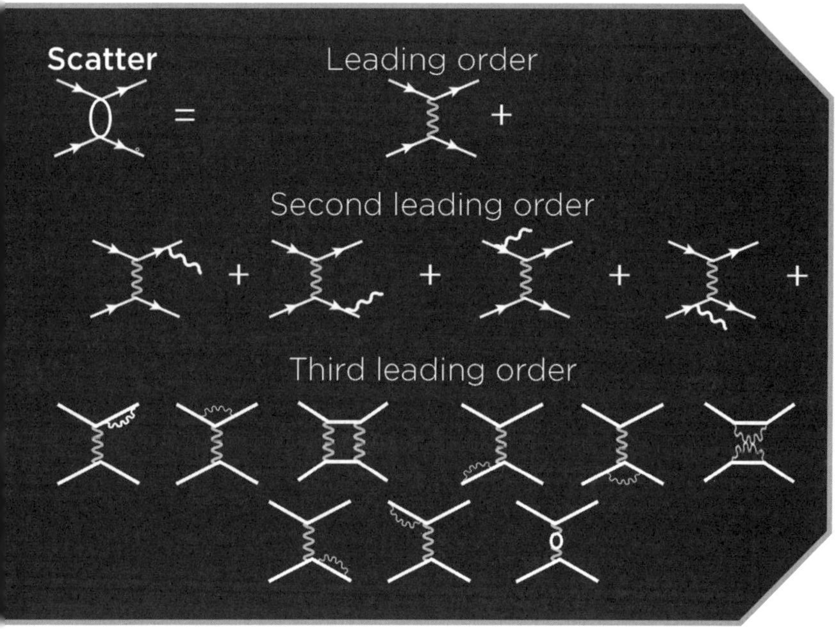

- It turns out that the probability of any one of the doodles depends heavily on the number of emissions or absorptions. Particle physicists use the word "vertex" to describe either an emission or absorption because it is often drawn with straight lines that come together. The more vertices, the less likely they are to occur. In fact, each vertex means that that particular doodle happens 1% of the time as often as a doodle with 1 fewer vertices.

- Given such a big difference between the different number of vertices, we can use perturbation theory to vastly simplify the situation. The full calculation is done by adding up all the contributions. If you want to calculate what processes matter when scattering electrons, it's really just the simple case that matters most, with 2 electrons in, 2 electrons out, and a single photon exchange—a single doodle. If you want to calculate things with even more precision, you need to include more doodles.

FEYNMAN DIAGRAMS

- While there were many people who helped develop modern QED, 3 people were considered to have made the most important contributions: Richard Feynman, Julian Schwinger, and Sin-Itiro Tomanaga. Together, they shared the 1965 Nobel Prize in Physics.

- Feynman invented Feynman diagrams. Interestingly, each Feynman diagram is really just an equation in disguise.

- The most important Feynman diagram is the one with 2 electrons in, 2 out, and a single photon exchange. It has 7 components: the 2 electrons coming in and 2 coming out, the photon, and the 2 vertices (the emission and absorption points).

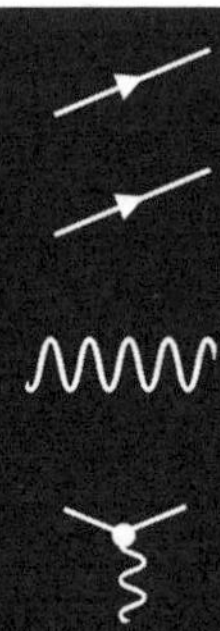

Object	Diagram	Symbol
Incoming particle		I
Outgoing particle		O
Photon		$\frac{-ig_{\mu\nu}}{p^2}$
Vertex		$ie\gamma^{\mu}$
Antimatter particle		I or O

- Each of these 7 components has a corresponding equation equivalent. For example, the incoming electron is represented by the symbol I, for incoming. Because there are 2 electrons, we can call them 1 and 2, so we have the symbols I_1 and I_2. The outgoing electrons we can denote with the symbol O, so we have O_1 and O_2. The other components have a more complicated set of symbols.

- To turn the simple and easy-to-draw Feynman diagram into an equation, all you have to do is take the symbols and put them together. This means that the simplest Feynman diagram corresponds to this equation:

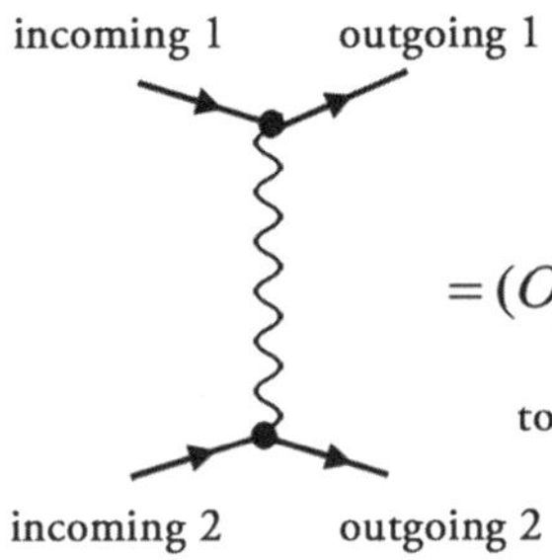

$$= \underbrace{(O_1 ie\gamma^{\mu} I_1)}_{\text{top electron}} \overbrace{\left(\frac{-ig_{\mu\nu}}{p^2}\right)}^{\text{photon}} \underbrace{(O_2 ie\gamma^{\nu} I_2)}_{\text{bottom electron}}$$

- In more complicated Feynman diagrams of even more complicated things, each of the squiggles, lines, loops, and everything else in those diagrams have corresponding mathematical symbols. So, if you know the rules, you can simply draw all possible diagrams, and the equations come essentially for free.

- But keep in mind that just writing down the formula isn't the same as solving it. Actual solving is difficult. The combination of Feynman diagrams to set up the equations and perturbation theory to simplify the solutions is what makes this field possible at all.

READINGS

Close, *The Infinity Puzzle.*
Farmelo, *The Strangest Man.*
Feynman, *QED.*
Pais and Jacob, *Paul Dirac.*
Schweber, *QED and the Men Who Made It.*

QUESTIONS

1. What are the key differences between somewhat familiar quantum mechanics and quantum electrodynamics?

2. How do Feynman diagrams represent real physical phenomena? In what way are they approximations?

NEUTRINOS VIOLATING PARITY AND THE WEAK FORCE

6

In 1898, New Zealand physicist Ernest Rutherford was investigating radioactivity and found that uranium seemed to consist of at least 2 types of radiation, which he named alpha and beta. Alpha radiation didn't seem to penetrate matter very much, while beta did. We now know that alpha radiation is the emission of helium nuclei (a composite particle consisting of 2 protons and 2 neutrons) and beta radiation is the emission of an electron. Alpha emission is governed by the strong nuclear force. Beta radiation is governed by the weak nuclear force—and this is the topic of this lecture.

WEAK FORCE

- Of the known subatomic forces, the weak force is by far the weakest. The decay times of radioactive decays caused by the weak force tend to be longer—many minutes, days, years, or even longer.

- Beta decay can change the identity of the substance doing the decaying. That isn't unique to beta decay. When uranium decays via alpha decay into thorium, it is changing the identity of uranium.

- But what is going on in alpha decay is very clear. Uranium 238 has 92 protons and 146 neutrons. The 92 and the 146 add up to 238. Then, it emits an alpha particle; it gives off 2 protons and 2 neutrons. So, it's left with 90 protons and 144 neutrons, which is thorium 234. But nothing really disappears; the protons and neutrons in the alpha particle are just shot off into space.

- But beta decay is different. For example, carbon 14, which is used to determine how old things are, decays into nitrogen 14 plus an electron. Carbon 14 consists of 6 protons and 8 neutrons, while nitrogen 14 consists of 7 protons and 7 neutrons. They both have 14 particles in them, so it seems that nothing is escaping the nucleus except the electron. What seems to be happening in this decay is that a neutron is transmuting into a proton.

- The weak force is unique among the different kinds of radioactive decay because it alone can change the identity of a particle, just by removing an electron.

- The weak force is different, and that matters every time we think about a grand unified theory. If the weak force can change particles' identities, but the other forces can't, figuring out how to reconcile the weak force with the others will be tricky.

- Another peculiar aspect of beta decays is that they don't even seem to conserve energy. The law of conservation of energy is one of the core principles of physics, and it has been since the early 18th century, when it was first proposed by French scholar Émilie du Châtelet. But when we measure the energy carried by the electrons from beta decay of carbon 14, we find that energy is missing.

ÉMILIE DU CHÂTELET
1706–1749

- There was another issue with beta decay that had to do with subatomic spin of the various particles. Protons, neutrons, and electrons all have either a spin of +½ or −½. Because both carbon 17 and nitrogen 14 have an even number of protons and neutrons, they must both have an integer amount of spin. The electron also has a spin of ½, which means that when the carbon 14 decays into nitrogen 14 and an electron, an integer spin object is decaying into another integer spin object plus one with spin ½. That means that the amount of spin changes in the decay.

- Because spin is actually angular momentum and angular momentum is also one of those things that is conserved, this is a problem. Could it be that in the subatomic world energy and angular momentum were not conserved? That was considered to be a real possibility by Niels Bohr, one of the legendary architects of quantum mechanics.

NEUTRINOS

- This problem took some time to resolve, and it wasn't until 1930 that the beta decay problem was figured out. As proposed by Austrian physicist Wolfgang Pauli, the neutrino was an electrically neutral particle that experienced the weak nuclear force and had very little mass. It also had a quantum mechanical spin the same as protons, neutrons, and electrons. Finding such a particle would be very challenging, and didn't happen for almost 2 more decades.

- In the meantime, Enrico Fermi published a theory in 1933 of beta decay, which gave essentially the first formal theory of weak interactions. Fermi used Pauli's properties for a neutrino and stated that a beta decay occurred when English physicist James Chadwick's newly discovered neutron turned into a proton and, in so doing, simultaneously emitted both an electron and a neutrino.

- Fermi's theory is basically what we think even today, with some minor changes. One change is that Fermi's theory was mathematically consistent only up until an energy of about 100 billion electron volts of energy. Above that energy, the theory gave nonsensical results. We now know how to solve the issues in Fermi's theory.

FREDERICK REINES (1918–1998) AND CLYDE COWAN (1919–1974)

- To verify that the weak nuclear force was described by Fermi's theory, the neutrino had to be discovered. In 1951, American physicists Frederick Reines and Clyde Cowan set out to find the elusive neutrino. And that's a very difficult thing to do. The weak force is weak, and neutrinos don't interact very much.

- Given the realities of neutrino detection, the trick is to get gobs and gobs of neutrinos and shoot them through your detector. Most of them will fly through the detector completely undetected, but a very tiny few might interact and be detected.

- Because neutrinos come from radioactive decay and other nuclear processes, Reines and Cowan put their detector very close to a working nuclear reactor, which irradiated their detector with 50 trillion antineutrinos every second on every square centimeter. Their target was about 200 liters of water containing the active detector material.

- Roughly, a few hundred quadrillion antineutrinos were going through their detector every second. After running for about 5 months, they saw 2 or 3 antineutrino interactions every hour. However, they could turn the reactor on and off and saw 5 times as many events per hour with it on as they did with the reactor off.

- On June 14, 1956, they sent a telegram to Wolfgang Pauli letting him know that his long-predicted neutrino had been discovered. Furthermore, the observed rate was within 5% of the prediction, thus validating Enrico Fermi's theory of weak interactions.

PARITY

- One of the questions about the known fundamental forces is whether they favor any particular class of particles or if they treat all particles the same.

- For example, the electromagnetic force treats matter particles and antimatter particles equally. In fact, as far as electromagnetism is concerned, you could swap the meaning of matter and antimatter—meaning that tomorrow an organization of physicists could simply decide that what we now call the positron is really the matter particle and the electron is the antimatter particle and the equations for electromagnetism wouldn't need to be changed. This is true for most of the forces.

- This idea that you can change something big, such as the meaning of matter and antimatter, without making any substantial change to your theory is a huge idea for unifying modern physics and is central to building a grand unified theory.

- However, it is possible that some forces might prefer some configuration over others. A symmetry we were hoping to find may not always be possible. And then, what seems to be a very obscure, small point turns out to have paradigm-shifting consequences.

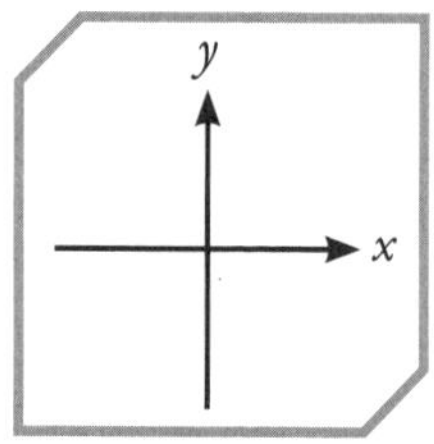

- Suppose that we draw a Cartesian plane, with an *x*-axis pointing to the right and a *y*-axis pointing upward. Would your physics equations care if you decided to flip the *x*-axis so it was pointing left and the *y*-axis so it was pointing down?

- The act of swapping left and right, up and down, and (3 dimensionally) forward and backward is called changing the parity of the situation. And if you make all of those changes and you can't tell the difference, your parity is said to be +1. If you make all of those changes and you get the opposite, the parity is –1. And if neither situation holds, the term "parity" doesn't apply.

- In the 1950s, scientists had started to study particle physics in detail. They looked at data generated from radiation hitting the Earth from space and had begun building powerful particle accelerators. These studies led to the discovery of many new particles, one of which was called a tau particle and another called a theta. Both of these particles had the same mass, but the tau particle decayed into 3 daughter particles, while the theta decayed into 2 daughter particles.

Eugene Wigner
1902–1995

- So far, there is nothing odd with any of this. However, when you considered parity, things became puzzling. If you added the equations involved with that 3-particle decay, the equations had negative parity, while the 2-particle decay had positive parity. This led to a completely different understanding of the weak force.

- In 1927, Hungarian physicist Eugene Wigner had studied the effect of parity on electromagnetism and found that, at least in electromagnetism, the universe didn't care if you swapped left and right—just like it didn't care if you swapped matter and antimatter. But a consequence of that observation was that parity should be conserved, meaning that the quantity should never change.

- There were 2 particles, identical in all respects, except that one decayed into a positive-parity state and the other decayed into a negative one. That meant that there were 2 kinds of parent particles that were identical except for parity, or that there was one kind of parent particle, but it didn't conserve parity when it decayed.

- This was a mystery, and in 1956, it led 2 Chinese physicists, Tsung-Dao Lee and Chen Ning Yang, to do an exhaustive literature search and realize that while the parity conservation of the electromagnetic force and the strong nuclear force had been experimentally verified, nobody had looked into it with the weak force.

- A colleague of theirs, Chinese-born Chien-Shiung Wu, a physics professor at Columbia University, was an expert in beta decay and decided to study a particular type of beta decay—specifically, when a nucleus of cobalt decayed into nickel, an electron, an antineutrino, and 2 photons.

Chien-Shiung Wu
1912–1997

- To do the experiment, she needed to be able to control the orientation of the subatomic spin of the cobalt nucleus. Spin in the subatomic world is complicated, but the easiest way to think about it in this context is to think of the cobalt nucleus as a spinning ball.

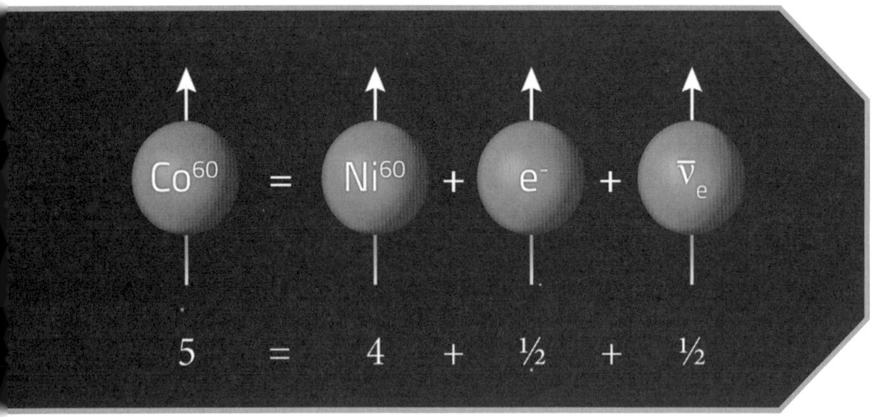

▲ The cobalt nucleus has a total spin of 5, while nickel has a spin of 4. The electron and neutron have a spin of ½ each, so it all works out: 5 = 4 + ½ + ½.

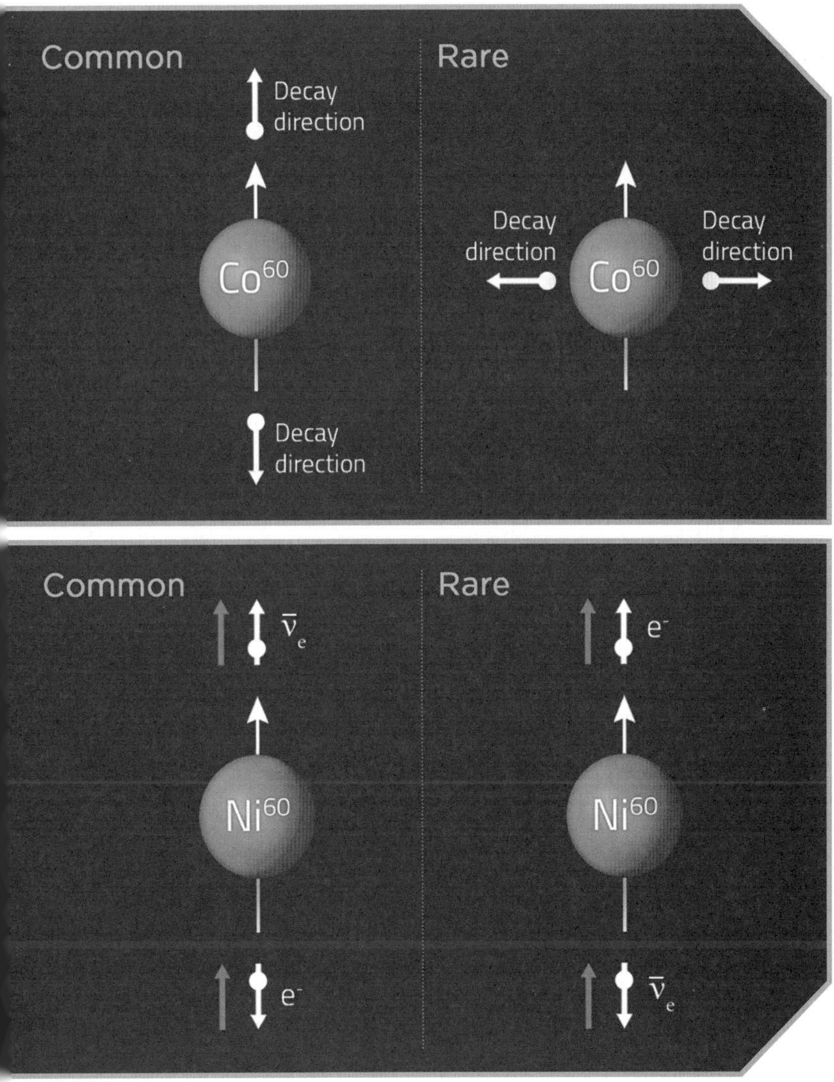

- Wu put the cobalt in a magnetic field to manipulate the spin of the cobalt atoms. She could do this because a spinning ball of charge like the nucleus makes a magnet, and magnets are affected by magnetic fields. She lined up the spin of the cobalt atoms. They were all pointing upward.

- After the decay, the nickel nuclei were also pointing upward. And, basically, the experiment was looking to see if the electrons decayed upward or downward. The expectation was that both directions should occur in equal numbers. That would imply that the equations conserved parity.

- However, Wu and her team found that the electron always decayed downward. This result shocked the community—yet Wu's experiment was proven to be correct.

- This meant that in weak-force decays, there was a clear preference for right or left. All antineutrinos spin one way and all neutrinos spin the other way. This means that the equations that describe the weak force care about left and right. There is a shorthand way to talk about this: All neutrinos are left-handed and all antineutrinos are right-handed.

- The weak force only interacts with left-handed particles and right-handed antiparticles. If our goal is to find a theory that unifies all the forces, our goal just got more difficult if we want to unify the electromagnetic force, which doesn't care about the handedness of the particles, and the weak force, which cares very much.

READINGS

Close, *The New Cosmic Onion*, chaps. 2, 8, and 11.
Lederman and Hill, *Beyond the God Particle.*
Lederman and Teresi, *The God Particle*, interlude C.
Lincoln, *Understanding the Universe*, chap. 4.

QUESTIONS

1 Why was the missing energy that led to the proposal of the neutrino so surprising?

2 Why was the manner in which the weak force interacts with particles of different handedness of spin so surprising?

7 FLAVOR CHANGES VIA THE WEAK FORCE

In the 1930s, some scientists had started to perform detailed studies of radiation from space—the rich bounty of particles pelting us from the cosmos. The basic idea was that this might teach us something about how the Sun burned and presumably how the hydrogen atoms that boiled off the Sun would interact in the Earth's atmosphere. That heralded a shift in scientific mindset, changing from big ideas, such as trying to find the building blocks of the universe, to seeing how it all fit together. However, the universe had a different plan.

TWO KINDS OF NEUTRINOS

- In 1932, American physicist Carl Anderson discovered a particle in his data that had the same mass as an electron but an opposite charge. He had discovered the positron, which is an antimatter form of the electron. The discovery of antimatter was already a big shock, but at least the positron was linked to the familiar electron. But in 1936, everything changed when Anderson discovered another particle in his data.

- While the electron had a very low mass and the proton and neutron had a high mass (about 2000 times heavier than the electron), this new particle was in between, with a mass of about 10% that of the proton.

- Two years earlier, Japanese physicist Hideki Yukawa had proposed the term "meson" to describe other theoretical particles in this mass range. Because Anderson's discovered particle didn't have the properties of Yukawa's proposed one, it got a slightly different name to distinguish them. Anderson's meson was called a mu-meson, or simply a muon.

- We now know that the muon is a heavy cousin of the electron, but that wasn't obvious for decades. And the meaning of the word "meson" has changed over the years, and we no longer consider the muon to be a meson.

- The muon isn't found in the atom and plays no role in nuclear physics or chemistry. It initially seemed to be unneeded. In fact, however, the discovery of the muon can reasonably be considered the birth of modern particle physics. Prior to that, the study of the subatomic realm was basically the study of the building blocks of the atom.

- From the 1930s through the 1950s, scientists studied their data from space—and, increasingly, data generated using particle accelerators. They found that they could make all kinds of particles that they could have never imagined. These particles had all kinds of masses and charges and spins.

- Of all the particles that they discovered, there was another meson that was of interest. In 1947, physicists at the University of Bristol in England had put detectors high in the French Pyrenees and found another particle. This one was called a pi meson, and it had a mass a little higher than the muon. This turned out to be Yukawa's meson.

- This particle decayed into a muon and an invisible particle that nobody could detect. This is very much like beta decay. In fact, it was pretty clear that the pion was decaying into a muon and a neutrino. And because a neutrino was involved, that meant that the weak force was coming into play.

Enrico Fermi
1901–1954

- Neutrinos are emitted in beta decay. The energy of the neutrinos involved in beta decay are relatively low, about a million electron volts.

- Enrico Fermi's theory of weak interactions was a great theory, but it was only an approximate model. Scientists knew it was only approximate because when you calculated what happened when a neutrino had a lot of energy—for example, 100,000 times higher than the nuclear realm—the theory predicted impossible things. Still, the theory worked pretty well at low energy, so people used it.

LEON LEDERMAN (B. 1922), JACK STEINBERGER (B. 1921)
AND MELVIN SCHWARTZ (1932–2006)

- On the other hand, because it was known that the theory was wrong at high energy, we needed a new source of data to better understand the behavior of the weak force at high energy.

- In 1962, American physicists Leon Lederman, Mel Schwartz, and Jack Steinberger used a particle accelerator to make a beam of pions. The energy of the pion beam was much higher than typical nuclear energies. And because pions decay into muons and neutrinos, this was a way to make a high-energy beam of neutrinos. This was the first beam of neutrinos ever made, and it meant that physicists could finally begin to explore the weak force at high energies.

- They knew that neutrinos came from beta decay, which were decays involving electrons, and from pion decay, which were decays involving muons. Decays occur when a heavy object turns into a lighter object. Using Einstein's equivalence of mass and energy, a high-energy object decaying into a lower energy object can emit neutrinos with energy left over. But this goes both ways, meaning that energy plus neutrinos can make a higher-mass object.

- They started with a high-energy beam of neutrinos that also had a lot of energy. If ordinary beta decay started with a neutron and ended up with a proton, a neutrino, an electron, and energy, then if you instead had a neutrino and energy, you could convert a proton into a neutron and an antimatter electron.

- And if you had enough energy, instead of an antimatter electron, you could make an antimatter muon. They had a beam with enough energy, so the idea was that they could slam a beam of neutrinos into a slab of steel and make electrons and muons. This would allow them to study the weak force at high energy.

- When they conducted their experiment, the first neutrino interaction made a muon, as did the second and third. In fact, after dozens and dozens of interactions, they never saw an interaction in which the neutrino beam made an electron. That was extremely weird. Neutrinos were particles involved in the weak force, and this particular beam of neutrinos only made muons when it hit a steel target.

- We now know that there are 2 distinct classes of neutrinos: an electron type and a muon type. We've also found a third type associated with a type of particle called a tau lepton, which is a heavier cousin of the electron and muon, but that wasn't relevant in 1962.

- There is more than one kind of neutrino, and this is another clue in the mystery of the weak force. The elusive and barely interacting neutrinos come in 3 distinct types, and these types are created with or create only their respective particle. All 3 types of neutrinos interact only via the weak nuclear force, and scientists now use beams of neutrinos to make precision measurements of that force.

NEUTRINOS CHANGE FLAVOR

- The existence of distinct neutrino types leads to an important concept in particle physics: flavor, which basically just means kind, type, or variety. The word "flavor" is used throughout particle physics to indicate things that are pretty similar yet distinct.
- Subatomic particles are divided into quarks and leptons. In the standard model, quarks are nuclear-like particles and charged leptons are electron-like. Quarks and leptons are pretty different, and accordingly, we would say that there are 6 flavors of quarks and 6 flavors of leptons, 3 of which are electrically charged and 3 of which are neutral.
- Flavor has an important place in our understanding of the subatomic realm, and the weak force has a unique role when it comes to flavor.
- An electron is an electron, and a muon is a muon, and the 2 particles are different. So, we make up a number that says that. An electron has an electron number of +1, and because it isn't a muon, it has a muon number of 0. Conversely, the muon has a muon number of +1 and an electron number of 0. This basic pattern is true of all of the matter particles—that is, all of the quarks and leptons.
- Antimatter particles are the opposite. For example, an antimatter electron, called a positron, has an electron number of –1.
- The force-carrying particles—the photon of electromagnetism, the gluon of the strong nuclear force, and the W and Z bosons of the weak force—don't have any flavor number.
- Neither electromagnetism nor the strong force can change a particle's flavor, but the weak force can.

- Photons can carry a lot of energy, and when energy converts into matter, it also makes an equal amount of antimatter. For example, a photon can convert into an electron and a positron. A photon has an electron number of zero. When it converts into the electron/positron pair, the electron has an electron number of +1 and the positron has an electron number of −1. Together, the 2 particles have a net electron number of zero—just like the photon.

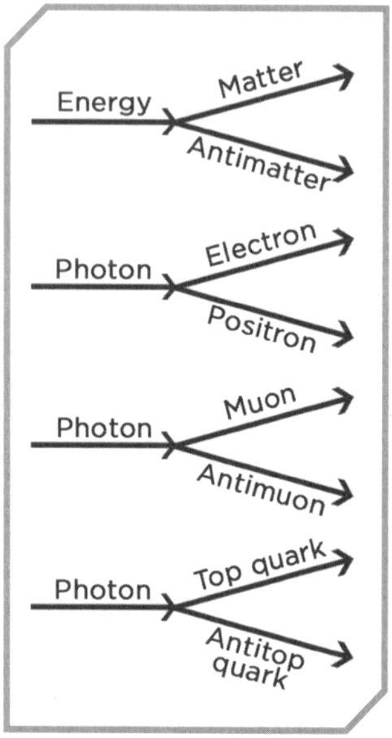

- This pattern is true of all subatomic particles. This is also conceptually true of the strong force, although electric charge isn't relevant there.

- A muon has a muon number of +1. If it decayed into a lighter particle—for example, an electron—the new particle would have an electron number of +1 and no muon number. Because the electromagnetic and the strong forces can't change a particle's flavor, that means that if all that existed were those 2 forces, the muon would be stable and never decay. The same goes for all of the other particles.

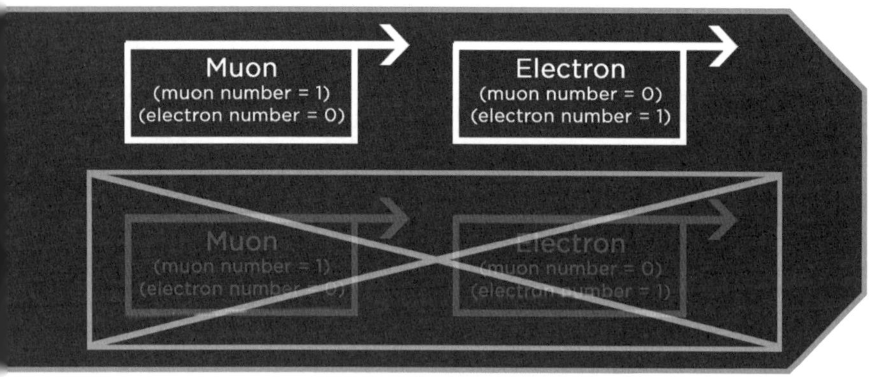

- But we know that the muon decays in about 2 millionths of a second, so something has to be going on—and that something is the weak force.

- Just as the photon mediates the electromagneti c force, the weak force is mediated by its own boson. In this situation, it's 3 bosons—specifically, 2 W bosons, one with a positive electrical charge and one with a negative one, and a Z boson, which is electrically neutral.

- The weak-force bosons are different than the photon. The photon is massless, which means that the electromagnetic force has infinite range. On the other hand, the weak-force particles are extremely heavy, which is why the weak force has a very short range. Another difference is that the W and Z bosons are unstable and decay very quickly.

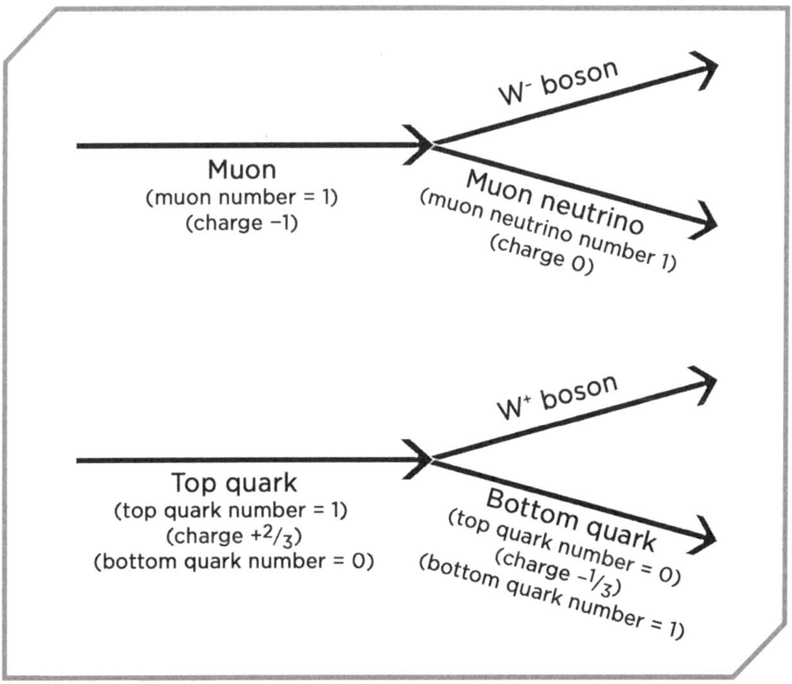

- When it decays, a matter muon, a muon with negative electric charge, simultaneously converts into a muon neutrino and emits a W particle, specifically the negative one. Before the decay, the electrical charge is –1, and after it is also –1. However, the muon number disappears, and a particle with a muon neutrino number appears. The flavor of the particle changes.

MASSIVE WEAK BOSONS

- A quark or lepton emits either a W boson or a Z boson. The W and Z bosons are the particles that mediate the weak nuclear force. They are analogous to the photon of electromagnetism. The Z boson is electrically neutral like the photon, but it is extremely heavy. The W particles are electrically charged and come in 2 versions: positive and negative. They are only slightly less massive.

- Only the W bosons are involved in changing the flavor of particles. If the Z boson could change flavors of particles, the top quark could presumably emit an electrically neutral Z boson and turn into a charm quark, which has the same charge as a top quark. However, this is never seen.

- The inability for the weak force to change the flavor of a particle without changing the charge was a mystery until 1970, when physicists Sheldon Glashow, John Iliopoulos, and Luciano Maiani figured it out. It turns out that if the quarks come in pairs, the weak force theory predicts that electrically neutral weak decays are forbidden by the theory. Even today, scientists look for instances in which the flavor of a quark was changed without changing the charge, because if we ever see such a thing, it will mean we've discovered something entirely new.

- A consequence of the mass of the W and Z bosons is the fact that the weak force proceeds very quickly at very high energy and very slowly at very low energy. Basically, at very high energy, the weak force is stronger (or at least can be stronger) than the strong force.

- This tells us that as we try to figure out a single unifying theory that underlies the existing theories, we need to take into account the fact that the relative strengths of the known forces change as we look at them at different energies.

- Once the decays of high-mass particles could be unified with the decays of low-mass particles into a single theory of the weak force, it also became possible to compare this unified weak force more directly with the other 2 fundamental quantum mechanical forces: the strong force and electromagnetism.

READINGS

Close, *The New Cosmic Onion*, chap. 8.
Lederman and Hill, *Beyond the God Particle*, chap. 10.
Lederman and Teresi, *The God Particle*, chap. 6.
Lincoln, *Understanding the Universe*, chap. 7.

QUESTIONS

1 Why was Lederman, Schwartz, and Steinberger's neutrino experiment a key discovery?

2 How does the high mass of the W and Z bosons make the weak force appear weak?

8 ELECTROWEAK UNIFICATION VIA THE HIGGS FIELD

With the discovery of the Higgs boson, the final missing piece of the standard model was found. More importantly, the effort to unify the weak and electromagnetic forces was complete. This step forward in the journey to devising a theory of everything was a story that took half a century to complete. This success in unifying 2 of the known fundamental forces is the biggest accomplishment in recent decades.

ELECTROWEAK UNIFICATION

- In 1960, American physicist Sheldon Glashow was toying with the weak and electromagnetic forces, when he devised a theory that was based on weak-force quantities. In this case, the term "weak force" is not being used in the same way that it's been used in the previous 2 lectures, which involved short distances and massive force-carrying particles. For Glashow, the term might be more accurately called the primordial weak force. This primordial weak force and the real weak force have some substantial commonalities.

- Glashow built a theory of this primordial weak force around 2 quantities: weak isospin and weak hypercharge. Isospin is a term that came from nuclear physics—the physics of the strong force. It's what holds protons and neutrons in the nucleus. While protons and neutrons have different electrical charge, as far as the strong force is concerned, the 2 particles are indistinguishable and you can swap them with impunity.

- Weak isospin relates to particles that the weak force can't distinguish. And if the weak force can't distinguish between the particles, they can be changed into one another without any penalty. This idea and the flavor-changing concept are inextricably linked.

- Weak hypercharge is kind of like electric charge, but it's just the weak charge. Weak hypercharge predicts one force particle that interacts with the weak charge.

- Glashow's idea predicted 4 particles that interacted via this primordial kind of weak charge: 3 called the B particles (specifically, B^+, B^0, and B^-) and the A particle. The B particles were the 3 isospin-changing particles and the A was the hypercharge particle. The plus and minus don't mean electrical charge; they mean the ability to change a particle's isospin, which means to change its identity.

- Those particles seem like the 4 particles we learned about when we discussed the photon of electromagnetism and the W^+, W^-, and Z bosons, but the 2 are different. For example, the A and B particles are all massless, which is quite different from the particles we've encountered so far.

STEVEN WEINBERG
B. 1933

- In about 1960 and 1961, American physicist Steven Weinberg and Pakistani physicist Abdus Salam had a series of conversations about how to convert ideas like Glashow's into something that agreed with the observed behavior of the electromagnetic and weak forces, but their work continued to come back to predicting massless particles.

ABDUS SALAM
1926–1996

- Massless particles, like the familiar photon, have an infinite range and can travel forever. This just wasn't observed, so it was clear that the idea needed work.

- In 1964, a series of papers were written to propose the existence of a new kind of energy field throughout the universe that could give mass to subatomic particles. These papers had 6 different authors, but the most famous of them is English physicist Peter Higgs, and we now call this energy field the Higgs field. A consequence of the Higgs field is that it also predicted that interactions in the Higgs field should be made possible by a new particle called the Higgs boson.

PETER HIGGS
B. 1929

- In 1967 and 1968, Salam and Weinberg independently took the ideas of Higgs and colleagues and applied them directly to Glashow's earlier ideas. The Higgs field, combined with the 4 particles of Glashow's theory, resulted in the forces of electromagnetism and the weak force. Glashow's ideas combining weak isospin and hypercharge are called electroweak theory, as it showed that both the weak force and electromagnetism had a common origin. Weinberg, Glashow, and Salam shared the 1979 Nobel Prize in Physics for their efforts.

- The biggest confirmation for the electroweak theory stemmed from a kind of weak force interaction that hadn't been seen before. The weak force was originally observed in beta decay, which is when a neutron emits a W particle and turns into a proton.

- The weak force was studied by American physicists Leon Lederman, Mel Schwartz, and Jack Steinberger in 1962, when they used a neutrino beam to hit nuclei and make muons. This was a case of the neutrino using its energy to emit a W boson and make a muon. In all cases, the weak force had been observed when a particle changed its identity, and this means that the W bosons were involved.

- However, an interaction involving the Z boson hadn't been observed. The Z boson interacts via the weak nuclear force and is electrically neutral. It is also massive. Making a Z boson is difficult, and finding it is even more difficult.

- You needed a beam of high-energy neutrinos that would emit a Z boson, which would interact with an atom. And because the Z boson doesn't change the flavor or charge of the particle, the neutrino would stay a neutrino, which means that you wouldn't detect it entering or leaving your detector.

- Because neutrinos don't interact very much and can pass through your detector without being detected, you basically needed an invisible and noninteracting beam that emitted a rare particle, and then the beam continued on without interacting.

- The only way you could see that this had happened was when an atom in your detector sort of blew up when it was hit by an invisible particle. This was a pretty difficult experiment to do. The scientific term for this kind of research is weak neutral currents.

- The detailed prediction of weak neutral currents was made by Glashow, Salam, and Weinberg in 1973, and their existence was discovered shortly thereafter in a neutrino beam at the CERN laboratory in Switzerland. Basically, a muon neutrino beam was shot into a large vat of liquid Freon, which allowed experimenters to see the debris of particles made in the neutrino interaction.

- Mostly, the neutrinos interactions that they detected were of the flavor-changing kind that made muons, but about 100 interactions were of the form where the neutrino entering and leaving the interaction were invisible and the atom just sort of blew up.

- Seeing atoms blow up established the existence of weak neutral currents, but neither the W nor Z boson had been observed yet. That would take another decade. At the CERN laboratory, a huge new particle accelerator had been built that would collide protons and antimatter protons together at extreme energies. While all kinds of interactions would occur, the clear goal was to make and identify the W and Z bosons. These were observed in 1983, and the discovery was awarded the 1984 Nobel Prize.

ELECTROWEAK SYMMETRY BREAKING

- Glashow started the process of figuring out how electromagnetism and the weak force originated in a more fundamental electroweak force when he invented a theory that involved the flavor-changing bits that originated in weak isospin and also in weak hypercharge. The problem is that while the theory predicted 4 particles just like we know arise in the real weak and electromagnetic force, those 4 predicted particles were all massless.

- This was a huge advantage for the theory because it put the weak force and electromagnetism on equal footing. Both forces predicted massless particles. Scientists have a word for this: symmetry. The symmetry in this case arises in the fact that they are all massless.

- On the other hand, we know that the W and Z bosons have mass, so we need another mechanism to give mass to some particles and not others. The technical term for this is to break the symmetry, and in this specific case, the term is electroweak symmetry breaking.

- The way to break the symmetrical situation of Glashow's theory into the nonsymmetric situation we see in our detectors—that is, the combination of massive particles of the weak force and massless photon of electromagnetism—is to postulate an energy field that we now call the Higgs field.

- The Higgs field can give some particles mass and not others. However, the Higgs field is not something you can actually see. But even if you can't see the Higgs field, there should be a particle making Higgs interactions possible: The existence of the field implies the existence of the Higgs boson, which is a particle that you can actually detect. And it ultimately was detected in 2012.

- The Higgs theory doesn't predict a single Higgs boson. It actually predicts that there are 4: Three of them are connected and called by the capital letter H with different superscripts—specifically, H^+, H^-, and H^0. The other one is alone and is labeled by a lowercase h.

- Recall that the names of the particles of Glashow's theory were the B^+, B^0, and B^- particles and the A particle. The B particles changed weak isospin and the A particle connected to weak hypercharge.

- The 4 H particles of the Higgs theory start pairing up with the Glashow particles. The H^+ particle combines with the B^+ particle and the result is the physical W^+. Physicists say that the Higgs particle "eats" the electroweak particles, but that's just a way of saying that they combine.

- The negative particles combine in the same way, with the H^- particle eating the B^- particle, resulting in the physical W^-.

- The combination producing the Z boson is a bit trickier. The B^0 and A particles first combine in a specific way, and this combined particle then is eaten by the H^0 particle. The photon is made by the B^0 and A particles combining in a second way.

- This combined particle doesn't interact with any of the 4 Higgs particles and consequently doesn't get mass. Thus, the physical particles that we can measure, the W and Z bosons and the photon, are a mix of Glashow's electroweak bosons and 3 of the 4 Higgs particles. That means that when the W and Z bosons were found back in 1983, scientists had already confirmed 3/4 of the Higgs theory.

- The final Higgs particle, the one denoted by a lowercase h, is the Higgs boson that was found in 2012. And with that discovery, the last missing piece of the standard model was found.

FINDING THE HIGGS BOSON

- People had been searching for the Higgs boson since it was proposed in the late 1960s, with experiment after experiment searching for it and coming up short. Each experiment ruled out a range of possible masses.

- By the time the experimental groups at Fermi National Accelerator Laboratory, known as Fermilab, turned their attention to the search for the Higgs, they knew that if the Higgs boson existed it had to be heavy—a bit more than 130 times heavier than the proton.

- There were many ways the Higgs could be discovered. Scientists at Fermilab converted the energy of the beams into a Higgs boson and looked to see it decay. The Higgs field gives particles mass, which means that the Higgs boson interacts more with heavy particles. That means it will decay into the heaviest particles it can. While the top quark and W and Z bosons are heavy, they are too heavy for Higgs bosons to decay into them. So, the next one in line is the bottom quark.

- The problem is that it was really easy to make bottom quarks at the Tevatron, which was the highest-energy particle accelerator in the world and was housed at Fermilab, so scientists needed to look for collisions in which Higgs bosons and either a W boson or Z boson were made at the same time. That was more difficult to do by ordinary interactions, and that meant that if they saw such a collision, it was much more likely to be a Higgs.

- Fermilab scientists took data for years, trying to accumulate enough to find a Higgs boson. In late 2011, the Tevatron experiments fired the first salvo by ruling out some possible masses of the Higgs boson. In the spring of 2012, the Tevatron experiments announced an update and were starting to rule out nearly all possible masses. The remaining window was small, and it was where the Higgs boson was most likely to be found.

- On July 4th, 2012, at a seminar at the CERN laboratory, the discovery of the Higgs boson was announced. In October 2013, the Nobel Prize committee called Peter Higgs and Francois Englert and told them that they had won the Nobel Prize for their 1964 prediction.

READINGS

Butterworth, *Most Wanted Particle.*
Carroll, *The Particle at the End of the Universe.*
Sample, *Massive.*
Lincoln, *The Large Hadron Collider*, chap. 6.
Randall, *Higgs Discovery.*

QUESTIONS

1 What analogy explanation for the way in which the Higgs boson works are you most comfortable with, and why?

2 How does the observed Higgs boson differ from the other 3?

QUARKS, COLOR, AND THE STRONG FORCE

9

You've learned how relativistic quantum mechanics, radioactivity, and the weak nuclear force are all tied together. There's one more force you need to understand to round out your understanding of modern subatomic physics: the strong nuclear force. This force is quite different from what you've learned so far, and to begin to understand it, this lecture returns to the early days of the 20th century.

NUCLEAR FORCE

- It was very clear not long after Rutherford's discovery that the atom consisted of a small nucleus consisting of positive charge, surrounded by a cloud of electrons. Initially, the nature of the nucleus was not well understood. In 1919, Rutherford discovered the proton. Rather than a nucleus that was a single ball of positive charge, it seemed to consist of a bunch of protons all stuck together.

- A nucleus of helium consists of 2 protons. The force they feel if they are touching one another is an appreciable repulsive force—about 20 pounds. In the case of the much larger uranium nucleus, a proton on the perimeter of the nucleus will feel a repulsive force of about 70 pounds.

- If the electromagnetic force of an atomic nucleus is pushing the constituents away with a force of tens or hundreds of pounds, there must be a force that is even stronger holding things together. And given the stability of the nuclei of most elements, that force is much stronger. That is why it is now called the strong nuclear force.

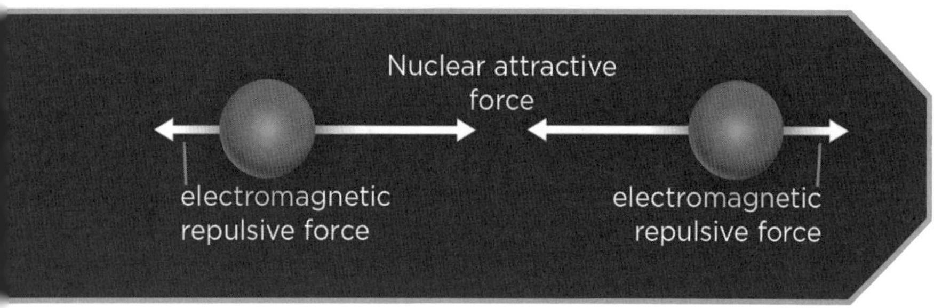

- We know that this nuclear force isn't the same as electromagnetism. For example, this strong force also holds the electrically neutral neutrons in the center of atoms, so there must be some sort of different kind of charge or interaction involved. In addition, we know that the range of the nuclear force has to be relatively small compared to electromagnetism.

- The reason we know this is there appears to be a maximum size of atomic nuclei. The trend is that atoms of larger atomic number tend to have shorter and shorter half-lives. And the fact that all nuclei heavier than about uranium naturally decay suggests that there's a point at which the nuclear force can't hold the nucleus together anymore. This means that the range of the nuclear force must be about a quadrillionth of a meter, called a femtometer.

- Just from these basic considerations, this new force must be much stronger than electromagnetism for sizes much smaller than a few femtometers and probably much weaker—perhaps even zero—for sizes bigger than that.

- The strong force seems to be a contact force. When 2 protons are far away enough that they don't touch, the strong force seems to not matter.

NUCLEAR FORCE MEETS COSMIC RAYS

- A quantum theory of the electromagnetic force led to the prediction of the photon as the quantum carrier of electromagnetism. In 1935, Japanese physicist Hideki Yukawa turned his attention to a quantum theory of the strong force and theorized that a quantum carrier of the strong force also existed. To explain the short range of the nuclear force, his particle had to have a mass. This was in stark contrast with the massless photon.

- Using theoretical considerations, Yukawa worked out the mass that the particle would have to have to explain the observed properties of the strong nuclear force. It would have to have a mass between that of the very light electron and very heavy proton. Yukawa called his proposed new particle the mesotron. It was later called the pi meson, or pion for short.

- If the first third of the 20th century was the period in which quantum mechanics and quantum electrodynamics (QED) was developed, the second third was a crazy time of discovery. People had invented new technologies to detect radiation and found that radioactivity was all around us.

- One of the most unexpected discoveries was that the Earth was constantly pummeled by radiation from space. The first discoveries of space radiation originated around 1910, but by the 1930s, new technologies gave scientists more tools to study the nature of this radiation from space. One of the early names for this phenomenon is cosmic rays.

- Starting in the mid-1930s, scientists were surprised to see all sorts of unexpected new kinds of particles created in their detectors. One of the first particles observed had the right mass to be Yukawa's mesotron, but it was quickly shown that the particle didn't experience the strong nuclear force, so it wasn't Yukawa's particle. This other particle is now called a muon.

- In 1947, an international trio of scientists working at the University of Bristol were studying cosmic-ray data when they found Yukawa's particle. Then, scientists all over the world started finding dozens of different kinds of particles in their cosmic-ray data. By 1951, Yukawa's particle was being called the pion, and more of the Greek alphabet was pulled in to give names to the new particles.

- There seemed to be many particles with masses in the midrange, and these were generically called mesons. But there turned out to also be many particles that had masses more like the proton and neutron and others that were even heavier. These heavier particles came to be generically called baryons.

- Most particles seemed to be created by the strong nuclear force and would decay via the strong nuclear force, although there were exceptions. One very unusual particle seemed to be created via the strong nuclear force but decayed very slowly via the weak nuclear force. Something was suppressing the ability of the strong force to let them decay. That was extremely strange, and this class of particles began to be called strange particles. There wasn't just one version of strange particles; whatever was causing strange particles seemed to affect an entire class of particles.

- The 1950s signaled the death of most of the cosmic-ray research industry and the birth of modern particle physics. Advances in detector and accelerator technology made it easy to make and study particles, and the discoveries rolled in. With such a glut of data, scientists of the era spent much of their time trying to classify the whole menagerie of particles, called the particle zoo.

QUARKS

- As for the strong nuclear force, an idea was proposed in 1964 that finally made sense of at least the strongly interacting particles that had been discovered in the previous few decades. American scientists Murray Gell-Mann and George Zweig independently proposed that they could explain the entire complicated mess of strongly interacting particles if they were composed of even smaller particles.

MURRAY GELL-MANN
B. 1929

- Gell-Mann called his proton constituents quarks and proposed that there were 3 different types of quarks, which are now called the up quark, the down quark, and the strange quark. If the electric charge of a proton is +1 and an electron is −1, up quarks have an electric charge of +⅔. In contrast, down and strange quarks have a charge of −⅓. The constituent quarks add together to form particles such as protons and neutrons.

GEORGE ZWEIG
B. 1937

- All quarks have a quantum mechanical spin of ½. They experience the strong force, and they combine in specific ways. One way quarks could combine is that a quark and antiquark could pair up and make what became the new definition for a meson, the middling mass particle. The other way quarks could combine was in groups of 3 quarks, and that became the new definition for a baryon—heavy particles such as protons.

- The quark model explained the many particles that had been discovered, although with different configurations. While the proton was an up up down configuration and the neutron was up down down, there was also down down down and up up up. There is also the down down strange, the down strange strange, and so on. There are 10 unique configurations of the 3 types of quarks, put into groups of 3.

QUARKS

uuu	ddd	sss	uud	uus
ddu	dds	uds	ssu	ssd

- Gell-Mann also assigned a mass to the quarks; more accurately, he just said that the strange quark was heavier. This was used to explain why the strange particles were heavier than the non-strange particles. And his model explained why the strange particles with 2 strange quarks were heavier still.

- When the quark model was proposed, the quark configuration with 3 strange quarks had not been discovered. For Gell-Mann's proposal to be right, a particle with 3 strange quarks would have an electric charge of –1 and a mass consistent with the 3-strange-quark hypothesis. The implications of his theory were clear, and the existence of a baryon with 3 strange quarks had to exist if he was right.

- In 1964, a group of scientists working at Brookhaven National Lab announced the discovery of a triple strange particle with exactly the properties predicted by Gell-Mann. This particle is now called the omega minus baryon.

- The quark model also explained the plethora of mesons. Mesons contained both quarks and antiquarks, and antiquarks had the opposite electric charge of their corresponding quark. There are 8 different ways you can select a quark and antiquark from 3 different flavors to make mesons.

- There is another oddity about quarks: We've never seen one. We've never been able to isolate and study a bare quark. They seem to be somehow trapped inside the baryons and mesons.

- In 1964, American physicist Oscar "Wally" Greenberg proposed that quarks had to have another property that neither Gell-Mann nor Zweig anticipated. He said that the 3 quarks had to have a property that is now called color. Specifically, there are 3 kinds of color: blue, red, and green. (The use of the word "color" has nothing to do with real color.)

- Because we knew that the proton and neutron didn't have any new and unexpected charge (or we would have observed it already), the 3 quarks in the proton or neutron each had to have one of the colors: red, blue, and green. Mesons are made of quarks and antiquarks, and antiquarks have anti-color.

- In 1974, a fourth quark was discovered, now called the charm quark, which has a charge of +⅔. In 1977, a new meson was discovered called the upsilon, and it was composed of a fifth quark, called the bottom quark, which had a charge of −⅓. A sixth quark, with a charge of +⅔ charge, called the top quark, was discovered in 1995. So, we know of 6 types of quarks, but there's no reason a priori why there couldn't be more.

QUANTUM CHROMODYNAMICS

- The modern theory of electromagnetism is called quantum electrodynamics (QED). In contrast, the theory of the strong force is called quantum chromodynamics (QCD). The "chromo" part is because the charge of the strong force is the color.

- In analogy with QED with its electric charges and photons that hold it all together, QCD had its color charges and a particle called a gluon, which transmits the strong force. Like the photon, the gluon is massless, and it has the same kind of quantum mechanical spin that the photon does. The gluon is also electrically neutral.

- There are, however, important differences between the photon and the gluon. The photon transmits the electromagnetic force but doesn't experience it itself. Because the photon is electrically neutral, it doesn't interact with other photons. Photons only interact with particles that have an electric charge. In contrast, the gluon does have color, which is the strong charge. Thus, gluons can interact with other gluons.

- This has other consequences for the stability of color for a quark. An electron is a negatively charged particle, and if it emits a photon, which is electrically neutral, the electron remains negatively charged. In contrast, when a blue quark emits a gluon, the gluon itself carries color, which means that the quark will change its color to either red or green. So, while each quark does have a color, its color isn't permanent—it's constantly changing.

- Gluons were discovered in 1979 at the DESY laboratory in Germany. But they weren't seen in the same way that photons were. Gluons, like quarks, experience confinement. Therefore, we don't see the bare quark in experiments.

- The mass of ordinary matter is a consequence of QCD. Most of the mass of humans isn't "stuff" in the sense we ordinarily think of it. Our mass is really just caused by the total energy stored inside the subatomic tempests inside of us called protons and neutrons. Our mass—indeed, all of the mass that we're familiar with—is just energy, governed by the laws of QCD.

READINGS

Griffiths, *Introduction to Elementary Particle Physics*, chaps. 2 and 8.
Lincoln, *Understanding the Universe*, chap. 4.
Veltman, *Facts and Mysteries in Elementary Particle Physics*, chap. 11.
Wilczek, *QCD Made Simple*.

QUESTIONS

1 What is the key difference between the gluon and photon that makes the strong force so qualitatively different from the electromagnetic force?

2 What consequences would there be for the nature of matter if the strong force were a long-range force like electromagnetism?

10 STANDARD MODEL TRIUMPHS AND CHALLENGES

At this point in the course, you are ready to bring it all together. Not only can you step back and see the complete picture of the physics that governs the subatomic world, but you can also begin to look for phenomena that are not yet explained and for which a newer and better theory is needed. This might offer a new and needed hint to help describe what a theory of everything might look like.

THE STANDARD MODEL

- In the standard model, the quarks are particles that only experience all 3 of the known subatomic forces: strong, weak, and electromagnetism. The leptons experience 2 of the 3, but they don't experience the strong force. The neutrinos experience 1, the weak force.

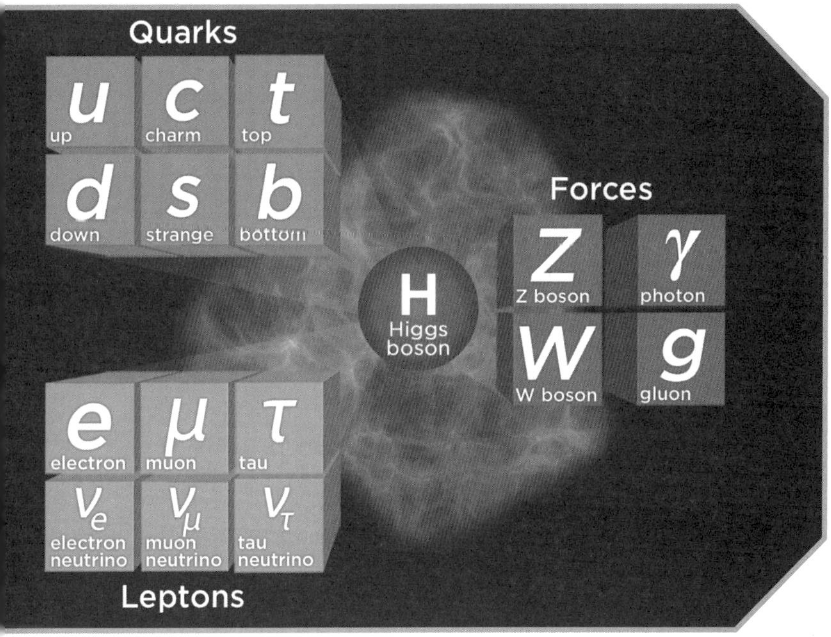

- All of the quarks and leptons together define a class of particles called fermions, which means that they are what can generically be called matter particles and they all have a subatomic spin of ½. Bosons are force-carrying particles with a subatomic spin that is an integer, usually 1 or 0. The photon, gluon, and W and Z bosons each have a spin of 1, while the Higgs boson has a spin of 0.

- The photon and gluon are massless particles, while the W and Z bosons interact with the Higgs field and have mass. The gluon is a bit peculiar in that it carries the strong charge and therefore interacts with other gluons. That gives the strong force a very different character compared to the other forces.

- The range for each of the forces is also very different. The electromagnetic force has an infinite range. The strong force is only strong, and only important, for distance scales about the size of a proton. The weak force only acts over a range about 1/1000th the size of a proton.

- These particles are the fundamental particles, which means that we don't know anything inside of them. But other names exist in particle physics that often involve combining the fundamental particles.

- It used to be that baryons, mesons, and leptons were classified according to the Greek roots of the words, with *barus* meaning "heavy," *mesos* meaning "middling," and *leptos* meaning "light." But the effort to classify particles merely based on mass reflected our ignorance of quarks. Our current nomenclature has baryons containing 3 quarks, mesons containing a quark/antimatter-quark pair, and leptons containing no quarks.

- What used to be called the mu-meson because of its middling mass does not have any quarks, so it's now referred to as a lepton. So, the muon particle we used to think of as a meson is still the muon, but it's now the mu-lepton.

- Ordinary matter is made of just 2 of the quarks—specifically, the up and down quarks found inside protons and neutrons and then the electron and its ghostly neutrino companion. The other quarks and leptons don't typically exist in ordinary matter; a particle accelerator or a high-energy proton from space crashing into the Earth's atmosphere is needed to make them.

- With the exception of the force of gravity, the standard model can explain the behavior of all known matter. It is a big step toward a theory of everything.

THE NEXT STEPS

- A theory of everything shouldn't have any holes. It should be able to answer all questions without any follow-up questions. The standard model doesn't live up to that very high standard.

- The biggest flaw is that gravity is not part of the theory. Perhaps a more manageable question deals with how the strong force fits into the electroweak picture. The electromagnetic and weak forces are consequences of a deeper and more fundamental theory, but the strong force isn't part of that.

- But even if we didn't aspire to the unification of all the 4 forces, simply within the context of the standard model, there are questions we can't answer. Ideally, the model would explain how everything is related.

- There are relationships between, for example, the mass of the top quark, the W and Z bosons, and the Higgs boson. If you are given 3 of the 4, you can tell what the third one will be, using the following equation.

$$\frac{m_W^2}{m_Z^2}(1-\sin^2\theta_W)=1+\frac{3G_F}{8\pi^2\sqrt{2}}m_t^2+\frac{\sqrt{2}G_F}{16\pi^2}m_t^2\,\frac{11}{3}\ln\,\frac{m_H^2}{m_W^2}\;+\cdots+\cdots$$

- However, there are quite a few parameters in the standard model that can't be calculated from first principles or other known parameters. These include such things as the mass of the quarks and leptons; the absolute strength of the strong, weak, and electromagnetic forces at a specific distance; and the likelihood that, for example, the top quark will decay into a bottom quark versus a strange quark versus a down quark.

- There are a few others that are a bit esoteric, but when you add them all up, there is a total of 19 parameters that we can't know in advance and that we currently need to simply measure in data and put in by hand. Once we have those parameters nailed down, we can, at least in principle, calculate the behavior of matter under all conditions.

- But having no theoretical basis to calculate the 19 parameters means that our theory is simply incomplete. There should be no such holes.

- Once we've realized that there is a grand mystery before us, we have to begin to think about what might be part of a better and more complete theory. To do that, we do 2 things simultaneously: constantly keep our eye on the final goal, which is a theory that answers all questions, and be on the lookout for unanswered questions. This second concern might point us in the right direction.

- For example, the fact that we can't predict the masses of the known fundamental particles means that we have mysteries in the realm of the Higgs field. After all, it is interactions with the Higgs field that gives particles their mass.

- Of the 19 unknown parameters, almost half of them are tied up in the masses of the quarks and charged leptons. We don't know why, for example, a bottom quark decays into a charm quark a certain fraction of the time and an up quark the rest of the time.

- We do know that the bottom quark has to decay into one or the other, so we know that the 2 fractions have to add to 100%, but why do we get the fractions that we measure? We simply don't know, but we do know that the force that changes flavor is the weak force, so that points us in that direction.

- There are some specific mysteries that are good prospects for where we might make progress in the near term. There are 2 questions that hopefully we might crack using the Large Hadron Collider, the most powerful particle accelerator to date. The grand hope is that if we can figure out a better and more inclusive theory that solves one of these questions, then that same theory will solve other mysteries. After all, that is the point of inventing a theory of everything.

WHY IS THE HIGGS MASS LOW?

- The Higgs field gives particles their mass; the Higgs field gives mass to the Higgs boson.

- While the Higgs boson can travel through space, it isn't just being a Higgs boson. Quantum mechanics basically says anything that can happen will happen if you wait long enough, and that means that the Higgs boson can temporarily turn into, for example, a bottom-quark/antiquark pair before converting back to a Higgs boson. Then, a little later, it might turn into a pair of W bosons or Z bosons.

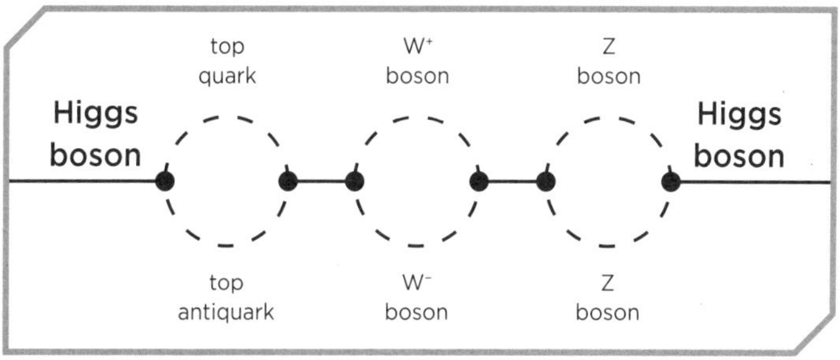

- The Higgs boson goes on its merry way, converting into this pair of particles and that. Perhaps even most bizarrely, it can temporarily convert into a pair of Higgs bosons.

- If you want to calculate the mass of the Higgs boson from first principles, you have to add up not only the contribution from the Higgs boson interacting with the Higgs field, but also the contribution from the fraction of time that the Higgs boson has temporarily turned itself into other particles.

- The measured mass of the Higgs boson should be the sum of the mass that we get when the Higgs boson is acting like a garden-variety Higgs boson and the mass when it has converted into something else.

> Mass[Higgs, observed] = Mass[Higgs acting like a Higgs] + Mass[Higgs turned into something else]

- The Higgs boson can convert into pairs of any particles that have mass, but it plays favorites. After all, it is through interactions with the Higgs field that some particles have more mass than others, and it prefers to turn into the heavier particles over the light ones.

- The big players are the top quark and the W, Z, and Higgs bosons. The others happen, but the contribution is pretty small. When we calculate the effect, we have to take into account the fact that the particles it can fluctuate into can occur up to the maximum energy that the standard model works at.

- The outcome of all these considerations means that the contribution to the mass of the Higgs boson due to the fraction of time when it has turned into something else is the product of the maximum energy for which the standard model applies and a function of the mass of the particles into which the Higgs boson can fluctuate.

- Because the Higgs, Z, and W particles are all bosons (meaning that they have integer spin) and the top quark is a fermion, which has spin of ½, we can compress the equation into the following form:

> Mass[Higgs turned into something else] = [constant] × [Maximum energy of the theory]^2 × [Mass(Higgs) + Mass(Z) + Mass(W) – Mass(top)]

- The mass of the Higgs boson is 125 billion electron volts (GeV), which is about 133 times heavier than the proton. The first equation said that if you use the standard model to calculate the mass of the Higgs boson from first principles, you get 2 terms. Because both are positive, this means that even if the first term were zero, this term due to morphing behavior of the Higgs boson can't be more than 125 GeV.

- The second term is a product of the maximum energy at which the theory applies (actually, the energy squared) and the mass difference between bosons and fermions. What is the highest energy at which the standard model is expected to apply?

- The theory should be valid all the way up to the Planck scale, which is about 10^{19} GeV. The real number is 10^{38}. Because the constant times the mass difference times 10^{38} can't be more than about 100, then either the constant or the mass difference has to be very small. And the constant doesn't save us. Thus, the difference in the masses of the fermions and bosons has to be small—in fact, almost precisely zero.

- In this calculation, we are forced to conclude that either the theory doesn't apply up to the Planck energy of 10^{19} GeV or there is an undiscovered rule that forces the fermions and bosons to exist on equal footing. There is a proposed theory that does that called supersymmetry.

- This calculation points in 1 of 2 directions: Either physics is just around the corner that invalidates the standard model at slightly higher energies than we study now, or we should investigate more thoroughly the linkages between matter and force-carrying particles (fermions and bosons).

WHERE DID THE ANTIMATTER GO?

- There is another unexplained mystery of the standard model that involves antimatter that might point us in a new direction.

- Antimatter is the opposite of ordinary matter. Combine matter and antimatter together and you get a huge amount of energy. The opposite is also true: Energy turns into matter and antimatter. A crucial point is that the amount of matter and antimatter must be exactly the same.

- The problem arises when we think about how this matter/antimatter equivalence jibes with what we know about the birth of the universe. Basically, we know that the universe began about 14 billion years ago in a cataclysmic event that we call the big bang. The universe was much smaller and hotter, and energy filled the universe.

- Given that we know a lot about this process from experiments, we understand with significant confidence that matter and antimatter existed in equal quantities in the early universe.

- But fast-forward to the present and everything around us is made of matter. Where is the antimatter? The short answer is that we don't know, but we do know a few things. A core premise of the standard model, which is strongly supported by measurements, is that we should either have equal amounts of matter and antimatter or, possibly, that the matter and antimatter could have all annihilated and the universe could be filled solely with energy. We see neither of these things.

- Combining observations and astrophysics knowledge, scientists have come to the conclusion that what seems to have happened is that during the early universe, before a tenth of a billionth of a second, some little matter/antimatter asymmetry developed. In fact, for every 10 billion antimatter particles that existed, there were 10 billion and 1 matter particles. The 10 billion pairs of matter and antimatter particles got together and annihilated, leaving just the 1 matter particle. And the descendants of that matter particle is the visible universe.

- This means that there must be some sort of new and unexplained physics that skewed the balance of matter and antimatter slightly. Investigating the weak force might be a promising place to start. In addition, deeper investigations into rare interactions mediated by the weak force might be a promising direction to pursue as we look toward a theory of everything.

READINGS

Close, *Antimatter.*
Giudice, *A Zeptospace Odyssey*, part 3.
Hoddeson, Brown, Riordan, and Dresden, *The Rise of the Standard Model.*
Lederman and Hill, *Beyond the God Particle*, chap. 12.
Lincoln, *The Large Hadron Collider*, chaps. 2 and 6.

QUESTIONS

1 Which of the pressing mysteries discussed in this lecture do you think will be solved first, and why?

2 How does the current standard model compare to the prevailing theory of the 1880s, when Newton's theory of gravity and Maxwell's equations were all that were needed to explain the phenomena known at the time?

11 HOW NEUTRINO IDENTITY OSCILLATES

The standard model is our current and best theory of everything, but it is incomplete. As you will learn in this lecture, the neutrino sector is full of clues about the next step toward a theory of everything, and over the next few decades, we may learn something crucial about neutrinos that will allow us make a big step forward toward a theory that explains everything.

THE SOLAR NEUTRINO PROBLEM

- In 1962, chemist Raymond Davis was interested in measuring neutrino emission from the Sun, and theoretical physicist John Bahcall was interested in calculating it.

- The Sun is an enormous and unregulated fusion reactor, combining hydrogen to create deuterium, then tritium, and finally helium. Along the way, neutrinos are emitted. The energy of neutrinos emitted in the process by which hydrogen is fused to be deuterium is very low and, given the technology available to Davis and Bahcall, was not amenable to detection on Earth.

- However, while the dominant energy-producing process in the Sun is the fusion of hydrogen to helium, that's not the only fusion process that occurs. Each process emits neutrinos of a different energy that can be detected by different technologies.

- In 1962, Davis had an idea to use a chlorine-based detector on Earth to search for neutrinos from solar processes involving beryllium. The neutrino would hit a chlorine nucleus and produce argon. Start with a pure sample of chlorine and look for the appearance of argon and you've done it. That's easy to say, but the reality was a bit daunting.

- The flux of neutrinos on Earth from the Sun is about 70 billion neutrinos per second through every square centimeter. Studies of the weak force had already shown that neutrinos don't interact very much, which means you need big detectors.

- Davis completed his detector in 1968 and would let his experiment run for several weeks while solar neutrinos turned chlorine into argon. The detector was big, the time was long, and the neutrino flux was enormous. Yet, given the low neutrino interaction probability, only a small number of neutrino interactions were expected.

- Every few weeks, Davis would bubble helium gas through the tank, hoping to see a few tens of atoms of argon—the number expected when he combined his detector acumen with Bahcall's predictions of the neutrino flux from the Sun.

- Davis found about a third as many neutrinos as predicted. Many tests were done, and both the experiment and prediction seemed solid. In fact, looking at these experiments from half a century later, we know that both men were right. The discrepancy is called the solar neutrino problem.

- There seemed to be something weird going on with neutrinos. Maybe there was some undiscovered error, so it was important to study neutrinos in another way that could confirm or refute the Davis/Bahcall result.

- Another way to probe neutrinos from a different source is from cosmic rays from outer space. Cosmic rays are generally just high-energy protons that hit the Earth's atmosphere.

- Even if you don't know how many cosmic rays hit the Earth's atmosphere, you know that cosmic rays are high-energy protons that decay through pions into muons, and you expect to see 2 muon-type neutrinos for every electron type. So, the ratios of neutrinos from space give a clear and unambiguous prediction.

- In 1985, a consortium of universities and national laboratories built another experiment called IMB to check on the neutrinos made by the collision of cosmic rays high in the atmosphere. The prediction was extremely robust: 2 muon types to 1 electron type. However, when the experiment was done, the detector saw equal numbers of muon neutrinos and electron neutrinos: The ratio was 1 to 1. This observation was called the atmospheric neutrino anomaly.

NEUTRINO OSCILLATION

- These 2 experiments, which were confirmed by many others, set the stage for a particle physics murder mystery. Where were the missing neutrinos going?

- There were many possible ideas kicked around, but the one that has survived is called neutrino oscillation. The idea is deeply rooted in quantum mechanics, but the basic idea is along the following lines.

- Suppose that there are 3 classes of neutrinos. Suppose further that these 3 types of neutrinos have mass, albeit a very low mass. It would be natural to think that these 3 types of neutrinos are just the electron, muon, and tau type, but that's not right.

- There are 3 different flavors (or types) of neutrinos, but they aren't the same as the 3 different mass neutrinos. Instead, each lepton flavor neutrino is a mix of the 3 mass types. And when you measure the mass of an electron-type neutrino, you will randomly find just 1 of the 3 possible masses, based on just 1 of the 3 mass types in the mix. Neutrinos don't have unique masses.

- The situation is complicated by the fact that we don't have instrumentation precise enough to measure the mass of neutrino types 1, 2, and 3. We only have instrumentation that can reveal their flavor.

- If this idea is true, it means that the 3 known flavors of neutrinos can morph into one another. Scientists call this process neutrino oscillation.

- If neutrino oscillation occurs, then it means that if you started with a beam of pure muon-type neutrinos and watched it for a long time, you'd see some of the muon types turn into electron types or tau types. These new types of neutrinos would then morph into different types. In the end, you'd see a beam containing 3 kinds of neutrinos, all constantly changing their identity.

- Practically, you can't run along a beam of neutrinos to observe that, but you can put detectors in the neutrino beam at various distances from the source and see the different neutrino populations.

- The oscillation of neutrinos depends on the differences between the masses of the 3 mass types, the energy of the beam, and—because the oscillation occurs over time and the neutrinos are traveling—the distance from the source.

- In this outlandish proposal, neutrinos don't have a fixed mass. This is different from other particles. In addition, the neutrinos morph from one kind into another. In fact, these 2 things are inextricably tied together: If you see neutrino morphing, it means that they have mass, and if they don't have mass, they can't oscillate.

- Neutrino oscillation would explain both the solar and atmospheric neutrino anomalies. The successful experiment to test it was brilliant in its elegance and simplicity. The first successful approach was to study the atmospheric neutrino anomaly.

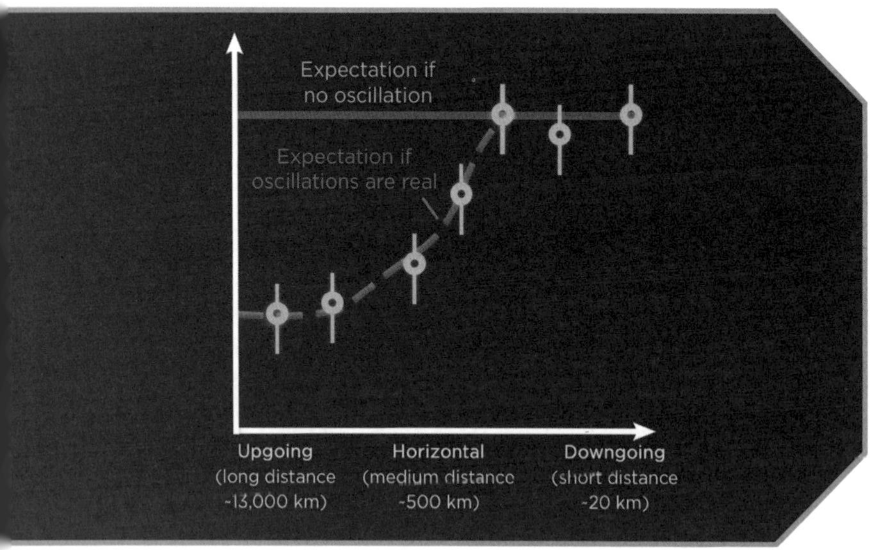

- The experiment that made the crucial observation is called the Super-Kamiokande experiment (Super-K), located in the Mozumi mine in Kamioka, Japan. The experiment showed unequivocally that when the expected ratio of muon to electron neutrinos of 2 to 1 was measured as 1 to 1, the reason was that muon neutrinos were disappearing. This is consistent with neutrino oscillation. The observation warranted the 2015 Nobel Prize in Physics.

- What were the neutrinos oscillating into? The distantly created muon neutrinos were disappearing, and there weren't more electron neutrinos. This implies that the muon neutrinos were turning into tau neutrinos, which the Super-K experiment couldn't identify properly.

- A few years later, in another experiment in Sudbury, Canada, called the Solar Neutrino Observatory, researchers looked at solar neutrinos and, for the first time, was able to see all 3 types of neutrinos: electron, muon, and tau. They proved that when they added up all the neutrino types, they saw the expected total neutrino flux from the Sun. And given that they could see all 3 types and other experiments couldn't, the theory of neutrino oscillations was established once and for all.

NEUTRINO MASS

- With the firm demonstration of neutrino oscillations, the fact of neutrino mass was also demonstrated. Neutrino oscillations do not mean that all neutrinos have mass; it means that the 3 types of mass-carrying neutrinos have different masses. Neutrino oscillations mean mass differences.

- For example, a neutrino of type 1 could have a mass of 0 while 2 and 3 have masses, or all 3 could have masses. But all we really know is that at least 2 types have mass. We know the difference of masses of the neutrinos, but we don't know their actual masses.

- If neutrinos are oscillating into other types of neutrinos, then you might wonder how many types of neutrinos exist.

- The Large Hadron Collider (LHC), currently the world's most powerful particle accelerator, is located just west of Geneva, Switzerland. The tunnel that houses the LHC once housed a different accelerator, called the Large Electron Positron (LEP).

- The LEP accelerator began operations in 1989 and was used to accelerate electrons and positrons and collide them together at the precise energy necessary to make enormous quantities of Z bosons, which is one of the particles that transmits the weak force.

- The LEP accelerator hosted 4 experiments in which Z bosons were studied extensively. Because Z bosons transmit the weak force and all particles experience the weak force, the Z boson decays into everything—quarks, charged leptons, and neutrinos.

- In 1989, LEP researchers published that the Z boson can decay into precisely 3 types of neutrinos—specifically, the electron, muon, and tau type. However, this simple statement can be made murkier.

- Before the 1998 definitive discovery of neutrino oscillations by the Super-K experiment, scientists were trying other ways to prove that neutrino oscillations were real. One such experiment was performed at the Los Alamos National Laboratory in New Mexico. They were generating a beam of muon neutrinos and looked for instances in which electron neutrinos appeared.

- In 1995, the experimental collaboration, whose name was LSND, found that they could most easily explain their data if there were 4 types of neutrinos. This announcement was not without controversy. The theoretical community imagined that there could, in principle, be 3 normal kinds of neutrinos and a fourth one that participated in neutrino oscillations but didn't experience the weak force. This new kind of neutrino is called a sterile neutrino.

- Sterile neutrinos were all the rage for a while. Some experiments found evidence for them while others did not. Recent analysis suggests that sterile neutrinos could exist with different parameters.

- We still don't know the answer. But the experimental controversy has resulted in a vibrant theoretical effort intended to work it all out. Because if we want to work toward a theory of everything, we need to solve this mystery.

- Under our current understanding of the Higgs field, which gives mass to the fundamental particles, particles either don't interact with the field or they interact enough that the outcome is a mass in the range of millions or billions of electron volts.

- It seems, at least as far as mass goes, that there are 3 different classes of particles: the quarks and charged leptons, with their millions and billions of electron volts of energy; the photons and gluons, with their 0 mass; and the neutrinos, with extremely tiny masses—masses that are close to 0, but not quite.

- We don't know why certain particles interact with the Higgs field and others don't. Suppose that neutrinos don't interact with the Higgs field, which is why they don't have a mass in the range of millions and billions of electron volts of energy. Maybe they experience a different interaction—maybe with a different kind of energy field or maybe some other mechanism—that gives them their tiny mass. Nobody knows if this idea is true, but it's possible, and it would explain the weirdness of the low mass of neutrinos.

READINGS

Close, *Neutrino.*
Jayawardhana, *Neutrino Hungers.*
Lincoln, *Understanding the Universe*, chap. 11.

QUESTIONS

1. Raymond Davis was able to extract about 10 atoms of argon gas. Given that perchloroethylene has about 6×10^{21} molecules per cubic centimeter and that the volume of perchloroethylene was 380 cubic meters, how many perchloroethylene molecules were there in his detector?

2. In the Super-Kamiokande detector, atmospheric neutrinos from above the detector appeared in the predicted ratio, while from below there was a deficit of muon neutrinos. Considering the shape of the Earth and the fact that these neutrinos come from the atmosphere, qualitatively, how do you expect the muon neutrino population coming from other places on the Earth would look?

CONSERVATION LAWS AND SYMMETRY: EMMY NOETHER

12

A huge component of working out a theory of everything is the concept of symmetry—an important, if underappreciated, idea from mathematics. This mathematical principle plays a key role in current physics theories, and it has helped us make enormous progress in our understanding of the universe. In this lecture, you will be introduced to this core idea.

SYMMETRY

- Symmetry, in a physics and mathematical sense, means that you can change something and nobody would notice. If you flip a picture of a human face so that you've swapped left and right, the picture looks pretty much the same. This means that the human face—indeed, the entire human body—has a symmetry to horizontal flips. The technical term for this is bilateral symmetry, and it is related to the property of parity.

- A flip symmetry can be expressed mathematically. In the x-y Cartesian plane, traditionally the x-axis points to the right. The right side of the plane is x, and the left side is $-x$. If you had a point 3 units to the right of the origin, you'd say that $x = 3$.

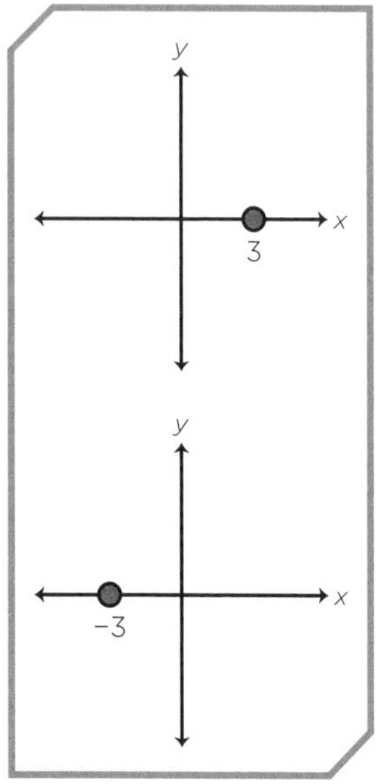

- If you swapped the axes and had the x-axis now pointing to the left, that would swap which side of the origin was positive territory and which was negative. Then, that point would now have a location of $x = -3$. Basically, if you simply replace all instances of x with $-x$, that's the same as swapping right and left.

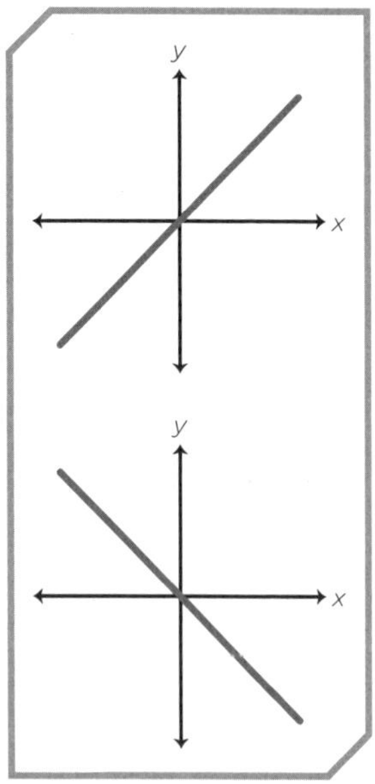

- In the simple equation $y = x$, if we replaced x with $-x$, we get $y = -x$. This equation after the swap is different from what we started with, which means that this equation is not symmetric under left-right swaps.

- In contrast, if you replace x with $-x$ in the equation $y = x^2$, we see $y = (-x)^2 = x^2$. In this case, when we square the equation, the 2 negative ones that are in front of the x multiply to a positive one, and the final equation is the same as the initial equation. Therefore, this equation is symmetric to left-right flips.

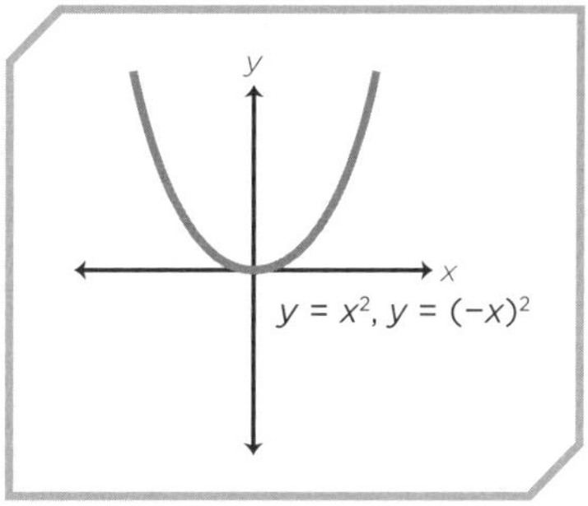

- In theoretical physics, if you can swap something or make a change in the equations that are describing the phenomenon that you are investigating and the outcome is it appears that you did nothing, then you say that the equations are symmetric under that operation.

- And if you somehow know that the equations have to be symmetric under some operation, that means that you can constrain the equations and rule in and out some terms. For example, suppose you know that the equation is symmetric under parity operations, which is the right-left flip. None of the graphs of the functions $y = x$, $y = x^3$, or $y = x^5$ are symmetric under that kind of flip.

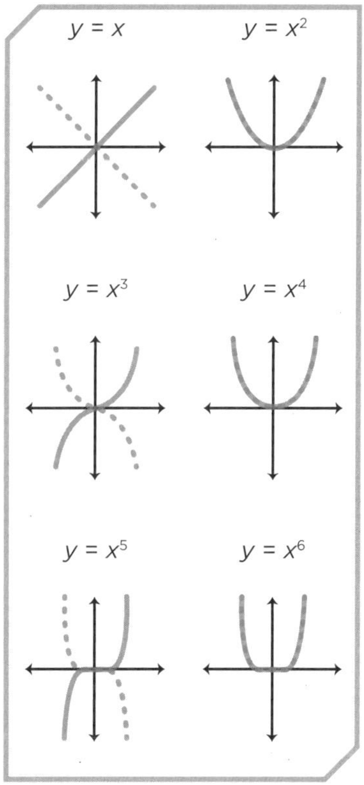

- In contrast, the $y = x^2$, $y = x^4$, and $y = x^6$ curves are all symmetric. Thus, if someone said that we knew that the physical theory had a right-left symmetry, then this would limit what the equations would look like. Solely using mathematical considerations, we could rule out the terms with odd numbers in the exponent by using the right-left symmetry requirement.

- This kind of symmetry turns out to be a huge component of working out a theory of everything. After all, the range of possible equations that can be written down is enormous. If we can use symmetry considerations to reduce the options, this provides an extremely powerful technique to help figure out a reasonable answer.

- There are classes of symmetries, such as discrete ones and continuous ones. A discrete symmetry is one like when you flip left and right—it's one or the other—or when you rotate an object by a fixed angle. A continuous symmetry is one in which you can make an arbitrarily small change, such as when you can rotate a circle by any angle, even tiny fractions of a degree, or when you can move an object by an incremental distance. These kinds of symmetries can have a real consequence on a physics theory.

EMMY NOETHER

- Amalie Emmy Noether was born on March 23, 1882. In a world in which women were not always appreciated for their intellect, she won the respect of Albert Einstein, David Hilbert, and Felix Klein, all of whom were world-famous researchers at the intersection of physics and mathematics.

- In 1915, Hilbert and Klein invited her to work at the University of Göttingen, a world-renowned mathematical institute, to help solve a key problem in the field of general relativity, which treats gravity as a bending of space caused by mass and energy.

- The theory seemed to have a serious sickness: Energy caused the bending of space, but gravity itself was energy. Thus, it would seem that the energy of bending space made yet more energy. Presumably, this would bend space more, resulting in more energy. It seemed like the theory could cascade in this manner to the point that energy would grow forever. And because this didn't happen, there needed to be a solution.

- Emmy's solution, which has come to be called Noether's theorem, was worked out in the same year she arrived and had far-ranging implications. There are many ways to write her theorem, and people have extended her initial work. Basically, the theorem says that if a set of equations that describes a physics situation has a continuous symmetry, then there is a corresponding conservation law—in which something is conserved if it doesn't change.

- Physics has laws of conservation of energy, momentum, and charge. And each of these conservation laws is associated with a symmetry in the equations that governs the laws of the universe.

- If we can move the position of an object without the laws of the universe changing, that leads to the law of conservation of linear momentum.

- Another symmetry of the laws of nature is that they are unchanged if we change what is defined as "day zero." The Gregorian calendar defines the first day as just a bit more than 2 millennia ago. Suppose that we defined the second at which the clock struck midnight on the night that split 1999 and the year 2000 as "moment zero." If the law of physics looked at that shift in time and yawned (and went back to sleep unchanged), then that means that the laws of physics are symmetric under shifts in time. The corresponding conservation law is the law of conservation of energy.

- If your physics laws don't care which direction you define as the "zero direction," this leads to the conservation of angular momentum.

- A symmetry that is a bit less intuitive is connected to the conservation of electrical charge. In quantum mechanics, the wave function can be used to calculate the probability that certain things happen and others don't.

- It turns out that you can shift the wave functions by an arbitrary phase. You can put in any arbitrary phase and the final answer doesn't change. Because the phase can be any value, this is a continuous symmetry. And, as Noether's theorem showed, each continuous symmetry corresponds to a specific conservation law.

- In this case, you can shift quantum mechanical wave functions by an arbitrary phase and this leads to the conservation of electrical charge. This phase shift is a more formal and slightly more complicated way to say that you can change the location you define to be at zero potential. The notion that conservation of charge is related to quantum mechanics is an awesome observation and an unexpected unification.

GAUGE SYMMETRIES

- There is one final type of symmetry that is a crucial component of modern theories: gauge symmetry. A gauge theory exists if you have a situation in which an experiment has an identical outcome in spite of having a very different definition of a baseline.

- An example of a gauge theory is time invariance. If you measure the amount of growth of a plant from today to tomorrow, you will get the same answer if you call today your "zero day" or a day 1000 years ago. The experiment will result in the same outcome no matter what you define as the zero day.

- Not all theories are gauge symmetries, but many are. In fact, all of the theories of the standard model and also Einstein's theory of general relativity are gauge theories. Gauge theories greatly constrain the equations of the laws of physics because it means that when you write down possible equations, they must have the property that the theory's predictions must be identical even if you somehow change the baseline or, equivalently, change what you define as your zero.

DISCRETE ABSTRACT SYMMETRIES

- While the continuous symmetries covered by Noether's theorem are extremely important, there also exist another class of symmetries that are both discrete and abstract.

- In old-fashioned nuclear physics, the forces that bind protons and neutrons to a nucleus are identical. You can swap all protons with all neutrons and, purely from a nuclear physics point of view, the nucleus will be identically stable. And a swap that makes no difference means that a symmetry is coming into play.

- Thus, in nuclear physics, an abstract symmetry arose when it was observed that the proton and neutron were 2 essentially identical states of what is called a nucleon. The nuclear force is a somewhat more complicated manifestation of the strong force, except that it binds nucleons together and not quarks and gluons together. And the nuclear force is blind to whether a nucleon was a proton or neutron.

- Because of that, a new property was invented called isotopic spin, or isospin, to keep track of whether the nuclear force is interacting with a proton or a neutron. The proton was given an isospin of +½, while the neutron had an isospin of –½.

- Isospin isn't a real thing; it's an abstract concept that involves an abstract symmetry. You can think of it as a bookkeeping trick to remember that there are 2 types of particles, even though the nuclear force treats both the same.

- The mathematical name for this particular symmetry is called SU(2) (special unitary group). This kind of isospin bookkeeping was later imported into electroweak theory, where this way of tracking symmetry helped make electroweak unification possible.

- Another example of an abstract symmetry appears in quantum chromodynamics (QCD)—specifically, color, or the strong nuclear charge. There are 3 so-called colors: red, blue, and green—which is arbitrary. Thus, you could swap green with red, red with blue, and blue with green and it wouldn't make any difference. The strong force would work just like it does now.

- The strong nuclear force and the theory of QCD are identical no matter how you define the 3 charges of the nuclear force. There is a symmetry for this, too, and it is called SU(3).

- There is a rather complicated mathematical definition to the groups SU(2) and SU(3), but basically, whenever we say that any group is SU(N), that means that there are N things that are indistinguishable under some particular change. The fact that the 3 colors of QCD are indistinguishable is called SU(3) symmetry. The 2 isospin values of +½ for the proton and –½ for the neutron are called SU(2) symmetry.

- Symmetries of this type have the potential to make significant predictions. For example, suppose that you observe some particular phenomenon and postulate that the phenomenon is governed under a particular abstract symmetry. That symmetry then imposes stringent requirements on the theory. One such example of the SU(3) symmetry is the quark theory when it was first proposed in 1964.

- Murray Gell-Mann's original theory only proposed 3 of the 6 quarks that we know now—specifically, up, down, and strange. In his theory, the 3 quarks are interchangeable. And a theory with 3 interchangeable things can be described by the SU(3) symmetry.

- Note that the interchangeability of the 3 quark colors and the interchangeability of the 3 lightest quarks are 2 distinct and unrelated examples of SU(3) symmetry. They have identical mathematical features but are very different physical phenomena, each with identical mathematical but distinct physical consequences.

- Abstract symmetries are mathematical rules that constrain theories but aren't theories themselves. The symmetries of early quark theory told researchers about the patterns for how quarks could combine, but symmetries were silent on topics such as how quarks actually interact.

- Abstract symmetries hint at real theories. And in our search for what comes next in the development of a theory of everything, these kinds of hints from symmetry can be very valuable.

READINGS

Griffiths, *Introduction to Elementary Particle Physics*, chaps. 4 and 10.
Lederman and Hill, *Symmetry and the Beautiful Universe.*
Livio, *The Equation That Couldn't Be Solved.*
Schumm, *Deep Down Things.*
Stewart, *Why Beauty Is Truth.*

QUESTIONS

1. How do symmetries lead to conservation laws?

2. Using the cat analogy described in the lecture, how is the example that all cats in all apartments fall equally a simple example of the much richer and more complicated idea of a gauge symmetry?

13 THEORETICAL SYMMETRIES AND MATHEMATICS

For anyone interested in trying to understand what's going on in the minds of theoretical physicists when they think about possible new theories of everything, symmetry is one mathematical idea that is necessary to understand. In the previous lecture, you learned about symmetries and their role in particle physics theories, with a concentration on symmetries of space and time. The lecture ended with a brief discussion of another class of abstract symmetries that are very helpful to understanding the laws of the universe: internal symmetries. This lecture will offer a peek at the mathematical underpinnings of these abstract ideas.

GROUPS

- The nature of internal symmetries is that they are based on purely mathematical symmetries called groups. While physicists are relative latecomers to the mathematics of symmetry, mathematicians have known about this kind of math for a very long time. Mathematicians call it group theory, and they understood the concepts quite well by the mid-1800s. Physicists reinvented some of these ideas.

- A group needs a set of elements and an operation. To illustrate, we can use the law of addition and integers. The set of elements is the integers, and the operation is addition.

- To be a group, it must have 4 properties:
  1. **Closure**. If you take any 2 elements and perform the operation, the answer must also be in the set. In the addition of integers, add any 2 integers and the result is another integer.

     integer + integer = integer

  2. **Associativity**. If you do the operation twice, you can choose the order in which to do it. This is the associative property of addition. For example, if you add 1 and 2 and 3, you can first add the 1 and 2 and then add the 3, or you can add the 2 and 3 and then add the 1. (Note that the commutative law of addition, which is if you add 1 + 2 you get the same thing if you add 2 + 1, is not a required property of a group.)

     $(1 + 2) + 3 = 1 + (2 + 3)$

  3. **Identity**. If you take the element and another element and do the operation, you end up with the original element. In the case of the addition of integers, this is zero, because if you add zero to anything, it doesn't change.

     integer + 0 = integer

4 **Inverse**. If you take 2 elements and do the operation, you get the identity element. In our example, these are the positive and negative integers. Add the same positive and negative number and you get zero. Add 5 and –5 and the result is zero.

> integer + (–integer) = 0

SUBATOMIC SPIN

- Groups that are interesting in particle physics can often be written in terms of matrices. There's no deep meaning in the ability to write theories as matrices, but it does make it helpful to call out quantities that are unchanged in the theory.

- The spin of fermions, which can be either positive or negative ½, has a mathematical structure that shows up repeatedly. This structure applies to all 2-component systems.

- The spin of fermions can be 1 of only 2 possibilities—specifically, + or – ½. You could represent them as 2 arrows, one up and one down.

> [↑↓]

- But it is mathematically convenient to write it as 2 columns that you can use matrix math to manipulate.

$$\begin{bmatrix} 1 \\ 0 \end{bmatrix} \text{ and } \begin{bmatrix} 0 \\ 1 \end{bmatrix}$$

- The top number is the positive ½ spin and the bottom is the negative ½ spin. If the top number is 1 and the bottom is 0, then the particle is 100% +½. If the top number is 0 and the bottom is 1, then the particle is 100% –½. Note that the columns don't say anything about directions, whereas the up and down arrows imply direction.

- That's one reason why the following method of representing spin is a bit better. This approach decouples spin and spatial directions and highlights the core mathematical ideas.

$$\begin{bmatrix} \text{up} \\ \text{down} \end{bmatrix} \text{ and } \begin{bmatrix} +\frac{1}{2} \\ -\frac{1}{2} \end{bmatrix}$$

- In quantum mechanics, the spin of an object doesn't have to be only up or down, except when we measure it. This is basically the same idea as Schrödinger's cat being alive and dead at the same time and only when we measure it do we know the outcome for sure. The spin of a fermion when you're not looking at it could be like this equation.

$$\begin{bmatrix} \alpha \\ \beta \end{bmatrix}$$

- The way you interpret alpha and beta is that the particle has a probability of alpha squared of being in the +½ spin state and a probability of beta squared of being in the –½ spin state. When you make the measurement, you find that the fermion is in one or the other of the 2 spin states. Alpha and beta just give you the probabilities of either outcome.

- Note that alpha and beta can be complex numbers, which means that they can include the square root of –1, which is written as *i*.

$$\sqrt{-1} = i$$

- That means that squaring them is a little trickier than the ordinary meaning of the word "squaring" a number, but that's just a wrinkle from the mathematics of complex numbers.

- Another tricky part is that when you measure the spin of an object, you measure it in a specific direction. We live in a 3-dimensional world, and we label those 3 dimensions as x, y, and z, which is arbitrary, but x could be left-right, y could be forward-backward, and z could be up-down.

- To measure the spin in, for example, the x direction, you need to have a matrix that pulls out the x information of the spin of the fermion. The same goes for y and z. The matrices that do this are called Pauli matrices, named after their developer, Wolfgang Pauli, who used them as one of the architects of quantum mechanics. His matrices are shown here with the identity matrix:

$$\sigma_x = \begin{pmatrix} 0 & 1 \\ 1 & 0 \end{pmatrix}, \quad \sigma_y = \begin{pmatrix} 0 & -i \\ i & 0 \end{pmatrix}, \quad \sigma_z = \begin{pmatrix} 1 & 0 \\ 0 & -1 \end{pmatrix}, \quad I = \begin{pmatrix} 1 & 0 \\ 0 & 1 \end{pmatrix}$$

- Matrices are important for the kinds of groups that are used in particle physics. In fact, if you know how to multiply matrices, you can multiply these same matrices together and see that they have lots of the group properties. For example, if you multiply the σ_x matrix by itself, you get the identity matrix. That's also true if you multiply σ_y by itself or σ_z. This particular mathematics not only describes spin but also protons and neutrons in nuclear theory.

- Suppose that we wanted to measure the spin of a fermion in any arbitrary direction. We can use these matrices to form a more complex matrix that can point in any direction. If you need to measure the spin in another direction, you can just rotate to that direction and the matrices automatically take care of the spin part. For that reason, the new matrix is called a rotation matrix.

- In this case, we're talking about measuring the 2-component spin in any arbitrary 3-dimensional direction, but that's a specific and special case. There are many other kinds of matrices, with more components.

SPECIAL GROUP NAMES

- Most of the matrices being discussed are rotation matrices and are associated with corresponding groups. The groups have names, and there is a code to the names. For example, a group with the name U is unitary. Although it has a complex mathematical definition, within the context of particle physics, unitary simply means that the matrices don't change the probabilities.

- This is an important factor when dealing with quantum mechanics, which lives and dies on probabilities. The groups also come with a number, which is the number of rows and columns in the relevant matrices. So, U(1) includes all of the unitary matrices with 1 row and 1 column, which is just a fancy way to say a single number. U(2) would be a unitary matrix with 2 rows and 2 columns.

- There is a special class of groups and matrices that have a special property. They have a determinant of 1. For a matrix with 2 columns and 2 rows, such as the following, the determinant is *a* times *d* minus *b* times *c*.

$$\det\begin{pmatrix} a & b \\ c & d \end{pmatrix} = \left(ad - bc\right).$$

- That's a mathematical definition for how the 2-by-2 determinant is calculated, but here's why the determinant is important: The intuitive meaning is that the area sketched out by the matrix is also 1. This also has the effect of ensuring that the probabilities of quantum mechanics still add up to 1.

- A matrix that is special in this way, where the probabilities add up to 1, has an S tacked in front of the rest of the group, so you might see something like SU(2), which is the geometry of the spin matrices discussed previously.

- The U matrices can include complex numbers. If you restrict yourself to real numbers, then the U is replaced by an O. For example, SO(3) is the group of all 3-dimensional rotations in the world of real numbers. Because SO(3) describes a 3-dimensional rotation, if a quantity is conserved under the SO(3) group, it means that it's the same in any direction, which means that angular momentum is conserved.

- SU(2) and SO(3) are very similar, although there is a tiny difference owing to the use of complex numbers versus just real ones.

- If you read anything about the standard model, you will likely encounter a phrase along the lines of "the standard model includes the U(1) × SU(2) × SU(3) symmetry group." We can break down this phrase into simple statements: U(1) is the group that covers electromagnetism, SU(2) is a 2-by-2 matrix that describes the weak force, and SU(3) is a 3-by-3 matrix that describes the strong force.

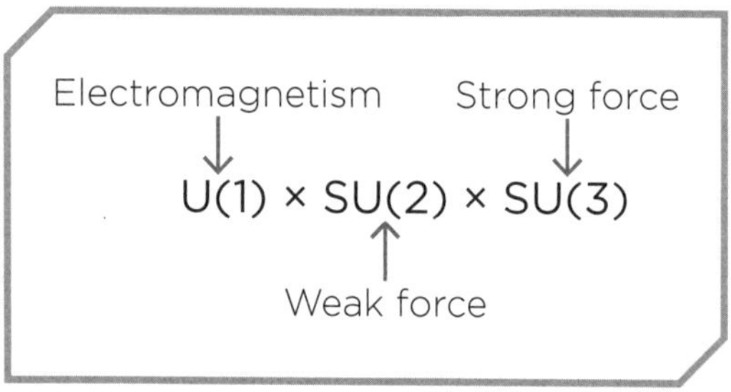

- While we now know how the symmetries and groups map onto the standard model, sometimes—especially for the strong force—the symmetries and associated groups came long before the actual theory was developed.

- This is likely to be the case for any future theory of everything. We will most likely start encountering new phenomena as we study data taken at higher and higher energy, perhaps new particles or perhaps new forces. It is almost certain that we will be horribly confused and that we will start working out the symmetries long before a fully developed explanatory theory is devised.

EARLIER ATTEMPTS AT THEORIES OF EVERYTHING

- If you have data to guide your thinking, using symmetries is pretty straightforward. But sometimes you don't, and that means that you speculate a little more freely. Speculating freely has a tendency to get scientists into intellectual trouble, but without data, theoretical physicists don't have much of a choice.

- One approach that they can take is to explore bigger symmetries that include the known symmetries within them. That way, you know that you are at least partly right.

- This is kind of like if you somehow represented the standard model as a square. A theorist could then speculate about a higher symmetry built on a cube. Because the standard model would be one face of the cube, that would already give you a leg up, and the other sides of the cube would point you to new phenomena to discover.

- SU(5) is the simplest symmetry group that can easily include the U(1) × SU(2) × SU(3) symmetries of the standard model. A theory in 5-dimensional space based on this was proposed in 1974 by Howard Georgi and Sheldon Glashow. The model predicts all of the standard model particles and 12 additional colored particles that cause protons to decay.

- Naturally, scientists tried to find examples of proton decay to verify the theory. There have been many attempts, but the theory is considered to be disproven. It is a grand and beautiful mathematical theory that simply doesn't represent the universe.

- The next simplest group theory to be considered is SO(12), which describes the rotations in 12-dimensional space. This theory contains SU(5) within it. This theory also predicts proton decay, which means that physicists don't like it much anymore. But it was a good idea. Howard Georgi also discovered this symmetry.

- There have been many theories proposed based on symmetries and group theories, but none of them have panned out yet. This doesn't necessarily mean that the approach is flawed; indeed, it is a very promising approach. But it hasn't worked out yet.

READINGS

Griffiths, *Introduction to Elementary Particle Physics*, chaps. 4 and 10.
Lederman and Hill, *Symmetry and the Beautiful Universe.*
Livio, *The Equation That Couldn't Be Solved.*
Schumm, *Deep Down Things.*
Stewart, *Why Beauty Is Truth.*

QUESTIONS

1. How does the mathematics of multiplication conform to the definition of a group?

2. What are the group properties of the geometrical examples of a triangle, square, pentagon, etc.? How do the group properties change as one continues to add sides and approaches a circle?

BALANCING FORCE AND MATTER: SUPERSYMMETRY

14

A new symmetry would lead us to new predictions and eventually to a more complete and newer theory of everything. There is one very popular symmetry candidate in theoretical physics circles called supersymmetry, which imagines that the matter particles and force-carrying particles are connected by a new kind of symmetry. There has been no experimental confirmation of supersymmetry thus far, but this approach has been considered by many scientists to be the most likely next big step in our journey toward a theory of everything.

SUPERSYMMETRY

- Supersymmetry, which often goes by the name SUSY for short, is a symmetry between force-carrying particles and matter particles. If SUSY is real, the key property that distinguishes between these 2 types of particles is spin.

- All the matter particles, both quarks and leptons, have fractional spin. All the force-carrying particles—the photon, gluon, and W and Z bosons—have integer spin. And SUSY is about trying to find a theory that doesn't care about spin.

- All of the force-carrying particles are of a class called bosons, which all have a subatomic spin that is an integral multiple of the Planck constant divided by 2π. This is called the reduced Planck constant, and the symbol we use is called *h*-bar, or $\hbar$. The spin of bosons could be 0 $\hbar$, 1 $\hbar$, 2 $\hbar$, etc. But because the $\hbar$ appears in all measures of spin, we drop it and instead focus on just the integer: 0, 1, 2, 3, etc. All of the bosons carrying 1 of the 3 traditional subatomic forces have a spin of 1, while the Higgs boson has a spin of 0.

- In contrast, all matter particles—all the quarks and leptons—are in a class of particles called fermions. All fermions have a spin that is a half-integer multiple of $\hbar$. For example, that means that fermions can have a spin of ½, 3⁄2, 5⁄2, etc. The quarks and leptons all have a spin of ½.

- Fermions and bosons are different in other ways, too. The rule for the matter-particle fermions is that no 2 fermions can be truly identical, meaning that they can't be in the same spot at the same time with the same spin. On the other hand, force-carrying identical bosons can be in the same place at the same time.

- Given that fermions and bosons act so fundamentally differently, it is difficult to imagine that any theory will treat them identically. Indeed, it is difficult to imagine why anyone would even think that they should be treated identically. Why do theorists believe that this might be something that the universe actually does?

- The laws of physics are symmetric under many operations, leading to many conservation laws. And in special relativity, the laws of physics need to be the same in all reference frames, meaning a conserved quantity and consequently a symmetry.

- The symmetry of special relativity is called a Lorentz symmetry, named after Dutch physicist Hendrik Lorentz, who developed the equations of special relativity before Einstein figured out the role the equations played in the nature of light.

- Because we learned from Emmy Noether's work that conservation of energy and momentum originate in symmetries of translation in time and space, we must add these to the equations. Combining Lorentz symmetries and space-time translations leads to the Poincaré symmetry.

- We can continue adding symmetries until we have added all of the possible symmetries that can exist and still have a mathematically consistent quantum field theory. It turns out that there is a small and finite number of symmetries possible. Moreover, all the known possible symmetries have been observed, except for one—between fermions and bosons.

WHAT DOES SUSY LOOK LIKE?

- A symmetry exists if you can make a change and that change makes no difference. In this context, that means that if you interchange fermions and bosons everywhere, the physical predictions of your theory are unchanged.

- Using f to denote fermions and b to denote bosons, suppose that the physics equation that governed the universe could be written as follows:

> Equation(before swap) $= f + b$

- If we swap fermions and bosons in the equation, the equation then turns into this:

> Equation(after swap) $= b + f$

- Because this is just simple arithmetic, we use the ordinary rules of the commutative property of arithmetic, and the swap makes no difference.

> Equation(before swap) = Equation(after swap)

- This equation has no physical significance, and we could have picked another example. For example, we could instead have chosen to multiply the letters f and b, and swapping the 2 makes no difference.

> Equation(before swap) $= f \times b$
> Equation(after swap) $= b \times f$
> Equation(before swap) = Equation(after swap)

- Both of these simple equations are symmetric and could be called supersymmetric. But not all equations are supersymmetric. For example:

> Equation(before swap) $= f - b$

- If we swap fermions and bosons, we get this:

> Equation(after swap) $= b - f$

- After the simplest algebra, the swap actually does make a difference for this equation. This particular equation isn't supersymmetric, even though the equation before and after the swap was pretty similar.

> Equation(before swap) = –Equation(after swap)

- If we wanted to do the same exercise in a manner that is more representative of the actual standard model equation, the equation would be much messier. The real equation is extremely complicated. But we can invent a stand-in. For example:

> Equation(before swap) = $f^2 + 4b$

- Swapping fermions and bosons results in this equation:

> Equation(after swap) = $b^2 + 4f$

- The equation before and after swapping are very different. The standard model and the standard model equation are not symmetric under swapping fermions and bosons.

- If we wanted to modify the standard model equation to be supersymmetric, we would have to change the equation. There are many ways to do that, but the easiest way would be to just add some terms.

- Let's keep the equation we used to stand in for the standard model. Temporarily and to make the new terms clear, let's use F and B for the new terms. To write a supersymmetric equation, we could write this one:

> Equation(SM stand-in) = $f^2 + 4b$
> Equation(supersymmetric SM) = $f^2 + 4b + B^2 + 4F$

- The point of supersymmetry is to make fermions and bosons indistinguishable, so we can rewrite that equation with all lowercase letters.

> Equation(supersymmetric SM) = $f^2 + 4b + b^2 + 4f$

- Then, if we swap the fermions and bosons, we get this equation, which is the same as the equation before swapping:

> Equation(supersymmetric SM, swap) = $b^2 + 4f + f^2 + 4b$

- We used equations because it's a convenient way to show what is going on, but it's important to remember that the fundamental and deep goal is to generate a theory that is symmetric if we interchange fermions and bosons.

- Even though we hope that supersymmetry might lead us to a new and important progress toward a theory of everything, it's important to remember that supersymmetry is not itself a theory. It is simply a principle that a supersymmetric theory must have. The principle gives general guidance, but only the theory actually tells us what to do.

- There are many theories that are supersymmetric. People can make all kinds of equations that include all kinds of physical ideas. This is one of the reasons why it is so difficult to prove, or eliminate, the idea of supersymmetry. You might rule out a particular supersymmetric theory, but to kill the entire idea is difficult.

SUPERSYMMETRIC COUSINS

- When we invented a stand-in supersymmetric theory, we just added some additional terms to the equation. That has to have some consequences, right?

- We know that the standard model works very well. It covers all the forces except gravity and it accommodates all the quarks, leptons, and force-carrying particles. If we add a mirror image of the equations, with the fermions and bosons swapped, that means that we have added to the equations a carbon copy of the particles of the standard model but with their spin changed.

- In the standard model, the quarks, leptons, and force-carrying bosons all must exist in a supersymmetric theory. But a supersymmetric theory must contain all of the supersymmetric cousins, or duplicate particles, too.

- The supersymmetric cousins of the fermions have the same name but with an "s" in front of them. For example, the quarks have the squarks. The system even works for the specific particles, with the selectron and stop squark. The entire class is called sfermions.

- For the bosons, the supersymmetric cousins have "–ino" added at the end of the word. The supersymmetric cousins of the photon, gluon, and W and Z bosons are photino, gluino, and Wino and Zino. The entire class of supersymmetry cousins for the bosons are called bosinos.

- The important thing about supersymmetry is that if you swap the fermions and bosons, you can't tell the difference. Thus, the sfermions, the supersymmetric cousins of the fermions, are all bosons, while the bosinos, the cousins of the standard model bosons, are all fermions.

- What exactly does that mean? The top quark is a fermion and it feels the strong, weak, and electromagnetic forces. It has a mass of 173 billion electron volts. The stop squark also experiences the strong, weak, and electromagnetic forces. In the simplest supersymmetric theory, it also has an identical mass of 173 billion electron volts. The only difference is that the stop squark is a boson. And that basic idea is true of all of the particles.

PROPERTY	TOP QUARK	STOP SQUARK
Feels strong force	Yes	Yes
Feels electromagnetic force	Yes	Yes
Feels weak force	Yes	Yes
Mass	173 GeV	173 GeV
Spin type	**Fermion**	**Boson**

- If the selectron is the same as an electron, but a boson instead of a fermion, then it has a mass of about half a million electron volts and experiences the electromagnetic force. Given that, we should easily be able to make selectrons in our accelerators. Yet selectrons have not been found.

- If the world truly exhibited supersymmetry, we would have found it by now. For example, matter and antimatter particles have exactly the same mass, so, with the exception of the proton, neutron, and electron, which we can find everywhere, we discovered matter and antimatter particles simultaneously.

- In spite of the fact that we've found all of the antimatter charge cousins, neatly arranged in the standard model, we have found precisely zero of the spin cousins we would expect from supersymmetry. So, the conclusion seems clear: The universe is not supersymmetric.

- However, we know of examples of broken symmetries. For example, the unification of the weak and electromagnetic forces was possible only because the symmetry is broken by the Higgs field. So, maybe supersymmetry is an example of a broken symmetry. Supersymmetry could be still true, but there could be some mechanism that makes the symmetry not quite perfect. The masses of the standard model and supersymmetric cousins could be different.

- In the case of supersymmetry, something must have happened that made the supersymmetric cousins heavier than the standard model ones. There are many ideas on this, with no real experimental guidance on what the truth is.

- The fact that scientists have looked for so long and not found the supersymmetric cousins has led many non-physicists to conclude that supersymmetry isn't real.

IS SUPERSYMMETRY DEAD?

- In the strictest sense, supersymmetry is dead. If supersymmetry were strictly true, selectrons would have the same mass as electrons and we'd have found them—and the same goes for all of the other cousins of the known standard model particles. But what about a broken symmetry, in which the supersymmetric particles exist but just happen to have a higher mass?

- It will never be possible to rule out a broken supersymmetry. No matter how high in energy experiments probe, it is possible for a true believer to say that supersymmetry is simply too beautiful of a mathematical concept to not be real and it's just that the cousins were more massive than we can find.

- We're left with the dichotomy of rejecting a strict supersymmetry and believing in supersymmetry that is broken ever more badly, with a larger and larger disparity between the masses of the standard model and supersymmetric cousins.

- Supersymmetry is the one missing symmetry that could exist in a mathematically complete and consistent quantum field theory that hasn't been seen. And it has potentially important consequences. Supersymmetry may be able to explain one of the most pressing problems still buried in the standard model: why the mass of the Higgs boson is low.

- Physicists also like the idea of supersymmetry because it can solve a mystery from astronomy in the gravitational realm known as dark matter, which is a conjecture invented to explain puzzling astronomical observations about the structure of our galaxy and other galaxies. From the perspective of particle physics, the dark matter conjecture states that there is a stable, electrically neutral, massive particle that can explain the astronomical observations. Supersymmetry provides a perfect candidate for such a particle.

- Supersymmetry is a very attractive principle for candidate theories offering a new model of reality, one that will subsume and replace the standard model. But it is not yet proven. So, research continues. In the end, data will answer this question.

READINGS

Hooper, *Nature's Blueprint.*

Kane, *Supersymmetry: Squarks, Photinos, and the Unveiling of the Ultimate Laws of Nature.*

———, *Supersymmetry: Unveiling the Ultimate Laws of Nature.*

Labelle, *Supersymmetry Demystified.*

QUESTIONS

1. Given what we know about the Higgs theory described in lecture 8, how many Higgs bosons do we expect in the simplest supersymmetric extension of the standard model?

2. How do the increasingly higher experimental limits on the discovery of supersymmetric particles affect the enthusiasm of the scientific community for the idea of supersymmetry, and why?

WHY QUARKS AND LEPTONS? 15

Symmetries are one way we can look for insights into the next discovery toward a theory of everything, but it's not the only way to find hints. We can also look at patterns that don't rise to the level of mathematical symmetries. One such pattern points us to the possibility that quarks and leptons might not be the smallest building blocks. Maybe they contain smaller particles still.

THE PERIODIC TABLE

- With regard to the periodic table of the elements, Nuclear physics explains the atom's mass, while quantum mechanics explains what the columns of the periodic table have in common. Specifically, the recurring electron configurations of quantum mechanics explain the chemical properties of the columns.

- The patterns of the periodic table hint hugely at what is going on, and once we know, they give us a compact expression of rules from both quantum mechanics and nuclear physics.

- In the standard model, we have the quarks (up, down, charm, strange, top, and bottom quarks) and the charged leptons (the electron, muon, and tau and their associated neutrinos). What kinds of periodic table–like patterns do we see?

- If the proton has a charge of +1, then the quarks in the first row—up, charm, and top—have a charge of +⅔. The quarks in the second row have a charge of −⅓. The charged leptons all have a charge of −1, while the neutrinos all have a charge of 0. So, in a way, the rows of the matter particles of the so-called periodic table of the standard model is analogous to the columns of the periodic table of atomic elements.

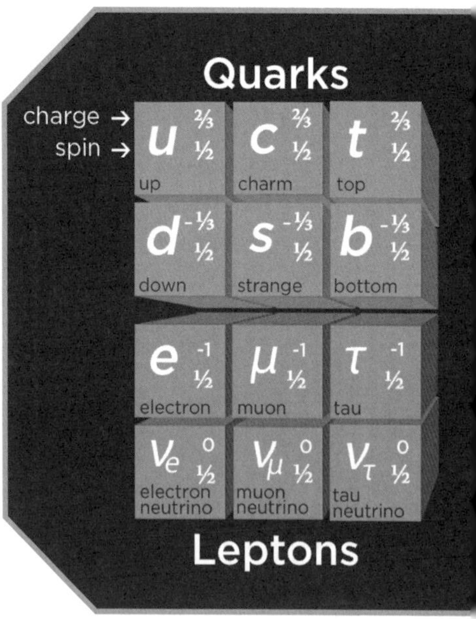

- Physicists call each column a generation. Generation 1 includes the up and down quark, the electron and the electron neutrino. Basically, generation 1 is what makes up the world we see around us. Generation 2 contains the charm and strange quark, muon and muon neutrino. Generation 3 contains the top and bottom quarks, tau, and tau neutrino.

- As we go across a row, from generation 1 to 2 to 3, the mass increases quite a bit as we go from left to right. For leptons, we see the same pattern. The mass of the neutrinos is much murkier, and we don't know the pattern very well.

- In the chemical periodic table, the chemically similar columns and increasing mass of Mendeleev's masterpiece as we moved from one row to another was fully explained when we finally understood quantum mechanics and nuclear physics. Likewise, there is some sort of explanation for the "chemically similar" rows of the periodic table of the standard model, with mass that increases as we move to the right.

- Nobody knows what the explanation for this is, but the general name for it is called the generation problem. Some people call it the flavor problem, because there are more flavors, or types, of quarks and leptons than we need to explain ordinary matter and also because quarks and leptons are different. But it boils down to the following: Why do generations 2 and 3 exist at all? And why are there quarks and leptons?

- The simple fact is that nobody knows. One explanation is that the quarks and leptons have smaller building blocks and that once we figure out the next step toward a theory of everything, the patterns we see will be as clear and obvious as the chemical periodic table. It's a viable theory, and one day it might be proven true, but for now, it's speculation and intuition.

- We can look at the data and ask what it tells us. The simplest way to address the question is to ask if quarks have a size, because if they have a size, then that almost certainly means that they have constituents.

- The size of a proton or a neutron is 10^{-15} meters, or a quadrillionth of a meter. So, how big is a quark? We can't answer that, but that's not the same as knowing nothing. We've looked at quarks and leptons and have tried to find their size, and the outcome of every measurement is what scientists call a null result. That means that we've never seen any inklings of a size.

- From that outcome, we can set an upper limit on the size of quarks and leptons. If they have any size at all, they must be smaller than about 1/20,000 the size of a proton, or about 5×10^{-20} meters.

- It might seem strange that we can draw that conclusion even if we've seen nothing. It boils down to knowing the equipment you're using. When we're looking to measure the size of quarks, we use the most powerful microscopes available, which currently is the Large Hadron Collider. The smallest thing it can see is about 5×10^{-20}. If quarks are smaller than that, we'll have to wait until the next big accelerator is built to see them.

PREONS

- The evidence for quark and lepton structure is pretty circumstantial and the experimental evidence is precisely zero, but that doesn't stop us from imagining what rules might govern the next deeper level might be as we try to find the true ultimate building blocks and the rules that govern them. In the absence of definitive data, we basically have to take the hints we have and guess.

- We can imagine what kinds of particles that could be found inside quarks and leptons, and it turns out that there are tons of ideas out there. The heyday of this type of theoretical work was the early 1980s, when theoretical physicists made many models. Each theorist came up with a different name for these hypothetical particles, but the generic term for the idea of quark and lepton constituents is preons.

- In 1979, Israeli physicist Haim Harari and American physicist Michael Shupe independently proposed a preon idea. They proposed that 2 types of preons existed: one with a charge of +⅓ and another with a charge of 0. Antimatter preons they proposed had the opposite electrical charge, −⅓ and 0.

- These preons are fermions, and each quark and lepton contains a unique mix of 3 matter or antimatter preons. Two preons of charge +⅓ and a preon of charge 0 made up the up quark. The positron (the antimatter electron) consisted of 3 preons with a charge of +⅓. We can identify the preon content of each of the particles of the first generation as follows:

CHARGE	PREON CONTENT	PARTICLE
+1	+ + +	e^+ [positron (antimatter electron)]
+2/3	+ + 0	u [up quark]
+1/3	+ 0 0	$\bar{d}$ [down antimatter quark]
0	0 0 0, $\bar{0}\,\bar{0}\,\bar{0}$	ν_e, $\bar{\nu}_e$ [electron neutrino and antineutrino]
−1/3	− $\bar{0}\,\bar{0}$	d [down quark]
−2/3	− − $\bar{0}$	$\bar{u}$ [up antimatter quark]
−1	− − −	e^- [electron]

▲ This choice gets all the charges right. The preon model also explains that preons are found inside the force-carrying particles—the photon, gluon, and W and Z bosons. The number of preons for the force-carrying particles is not 3; the weak force–carrying particles contain 6 preons, while for the electromagnetic particle, the photon, there are only 2 preons. The preon content of all of the force-carrying particles except the gluons are shown in this table:

CHARGE	PREON CONTENT	PARTICLE
+1	+ + + 0 0 0	W^+ [positive W boson]
−1	− − − $\overline{0}$ $\overline{0}$ $\overline{0}$	W^- [negative W boson]
0	+ + + − − − + + − − 0 $\overline{0}$ + − 0 0 $\overline{0}$ $\overline{0}$ 0 0 0 $\overline{0}$ $\overline{0}$ $\overline{0}$	Z^0 [neutral Z bosons] [There are 4 versions]
0	+ −	γ [photon]

▲ This all involves the first generation. How do the second and third generations fit in? In this particular model, the preon content of the second and third generations is not known.

- If we were vague about the preon content of the quarks and leptons of the second and third generations, we're very vague about the force holding together the preons. However, there is a name for this preon force and the charge that causes it: hypercolor. The basic idea is that the preons experience a hypercolor force that ties the preons together.

- The idea is that the known subatomic forces—the strong force, weak force, and electromagnetism—are just different manifestations of the hypercolor force. The dream is that hypercolor will explain all of the currently known forces except for gravity.

- Preons could be the ultimate building blocks of the universe. Of course, so could the quarks and leptons, which have never been shown to have smaller objects inside them.

- But it's also possible that the preons are just a stepping stone to the ultimate goal. If preons exist, there might be pre-preons and pre-pre-preons and a whole series of steps that lead down to the final and truly ultimate building block.

- Preons are not a popular idea; there are issues with the hypothesis. Perhaps the biggest one is a consequence of the Heisenberg uncertainty principle. Because preons are bound inside quarks and leptons and we know that the size of quarks and leptons are very small, then their location is highly constrained. The Heisenberg uncertainty principle says that a particle with a constrained location will have an uncertain energy, which is equivalent to mass. That means that the preons should have more mass than the quarks.

- That's potentially a problem, although perhaps a large negative binding energy will save the day. We really don't know. Frankly, there are issues with most hypotheses, so having issues alone is not fatal for a theory.

- There is more than one way to have constituents. After all, a competitor idea of the preon model is the concept of superstrings, which are thought to be the ultimate preon—maybe the pre-pre-pre-preon or something.

LEPTOQUARKS

- For now, preons are a pretty ambitious idea, and the jury is still out on whether they're real or not. But the basic question still remains: Why are there so many fundamental particles in the standard model? There are 6 quarks and 6 leptons. A theory of everything should originate in a single principle and maybe a single building block or maybe 2.

- Maybe we need to take a look at the quarks and leptons. Why should we have 2 different classes of particles like that?

- We know that quarks and leptons are different. The quarks feel the strong force and the leptons don't. But there's a deeper issue.

- Each particle flavor is assigned a kind of number. An up quark has an up-quark number of +1, while an antimatter up quark has an up-quark number of −1. Up quarks have a down-quark number of 0, a strange quark number of 0, and the same is true for all the other quark flavors. Up quarks also have a 0 for electron number, muon number, and so on. And the force-carrying bosons all have 0 for all the quark and lepton numbers.

- In strong and electromagnetic interactions, all quark and lepton flavor numbers are preserved. The weak force is different. It can change flavors. The numbers differ before and after decay.

- However, there is a deeper and even more general number that can be assigned to the quarks and leptons. We ignore the details of the flavors and say that any quark has a quark number of +1 and −1 for an antiquark. The same thing goes for the leptons, with leptons and antileptons having a lepton number of +1 and −1, respectively.

- Quark and lepton number seems to be religiously conserved in the standard model. There has never been observed a decay or a scatter or any process that changes net quark and lepton number. That means that a quark cannot decay solely into leptons. That would change the quark and lepton numbers.

- This is one way to distinguish quarks and leptons. But if we want to have a theory of everything, it would be helpful if we could somehow combine these 2 particles. This led scientists to hypothesize a hybrid particle called a leptoquark, which has both lepton and quark properties. It has both a quark number of +1 and a lepton number of +1.

- We haven't discovered leptoquarks, so that means that if they exist, they must be very heavy.

- Leptoquarks might be a unifying concept that is more basic and more fundamental than the familiar quarks and leptons. In addition, if they exist, leptoquarks could explain why there are equal numbers of quark families and lepton families; there is no reason why there should be the same numbers of both. Leptoquarks are a possible explanation.

READINGS

Kalman and Souza, *Preons.*
Lincoln, *The Inner Life of Quarks.*
———, *The Large Hadron Collider*, chap. 6.

QUESTIONS

1 Must preons be fermions or bosons, or could they be both? If preons could be only either a fermion or boson, but not both, which one is possible?

2 Do you find the qualitative argument of the similarities of the chemical periodic table and the properties of the quarks and leptons to be a compelling argument for the existence of quark and lepton substructure?

16 NEWTON'S GRAVITY UNIFIES EARTH AND SKY

We don't understand gravity in the same way as we do the other forces. Specifically, we don't have any idea how gravity works at the quantum level. This stems from the basic truth that gravity is incredibly weak. The other forces—even the weak force—so completely dominate gravity that we have no information on how it works in the realm of the super small. This lecture will explore what we actually do know about gravity.

ARISTOTLE

- In the 4th century B.C., Aristotle tried to explain gravity as being part of the nature of an object. This explanation came from a much broader picture of the cosmos, in which the Earth was at the center of the universe.

- Aristotle concisely described the early Greek's theory of celestial motion in a book called *Metaphysics*. The spherical Earth sat at the center of the universe. The other objects were embedded in concentric spheres surrounding the Earth, and each sphere moved because it was pushed by an individual deity.

- The first modern realization that the Sun and not the Earth was at the center of the universe was the epic book by Nicolaus Copernicus entitled *On the Revolutions of the Heavenly Spheres*, published in 1543. Copernicus's book retained the idea of spheres, but it set in motion a series of developments that led inexorably to the first success in our journey of the unification of forces. It was only a century after Copernicus's book that Newton explained how gravity on Earth and the motion of the universe came from a single source.

JOHANNES KEPLER

- Newton's triumphant unification of classical gravity built on insights from a German astronomer named Johannes Kepler and the contributions of Kepler's mentor, Tycho Brahe.

- Copernicus retained the idea of spheres, which was accompanied by the idea of circular orbits. The only problem was that the data disagreed with the simplest predictions of Copernican theory.

- Tycho Brahe was a Danish nobleman and astronomer who successfully argued that objects such as comets and what are now called supernovae were to be found at distances far from the Earth. This disagreed with Aristotelian thought, which claimed that transient astronomical phenomena must always be located closer to the Earth than the Moon.

- Brahe made what were, for the time, world-class astronomical measurements. His data, which included both his own observations and those of other luminaries of the European astronomical community with which he corresponded, was of excellent quality.

- Johannes Kepler, Brahe's assistant, applied the data to the topic of planetary orbits. He devised 3 distinct laws that at the time didn't seem to originate from any deeper theory and instead were just observations that were consistent with the data.
    1. The first law was that the motion of a planet followed an ellipse around the Sun with the Sun at one of the 2 foci of the ellipse. This was in contrast to the earlier ideas, in which the orbit of the planets was perfectly circular.

    2. His second law says that the area swept out by the motion of the planet is the same for equal units of time. And for this to be true, the planets must move faster when they are near the Sun and slower when they are far away.

    3. The third law says that the square of the amount of time to complete one orbit is proportional to the cube of the semimajor axis of the ellipse, which is the distance from the center of the ellipse to the perimeter in the direction that is farthest from the center. Basically, it's half the length of the long side of the ellipse.

- Kepler's laws were simply observationally derived and didn't originate in a deeper theory. In fact, one of the triumphs of modern gravitational theory is that Kepler's laws can now be predicted. The same is equally true of Einstein's more recent gravitational theory, which also predicts Kepler's laws. Kepler's first 2 laws were published in 1609 and his third was published in 1619, decades before Newton was born.

- There is perhaps one more early observation that needed a successful theory of gravity to explain. Galileo was an Italian scholar whose most important contributions to physics involved insights into the rate at which things fall when dropped.

- Aristotle's explanation of terrestrial gravity as being caused by objects finding their natural location suggested that heavy objects would fall faster than lighter ones. Galileo experimented with objects rolling down inclined planes and found that objects of the same size, but different masses, rolled with equal speeds.

- This nonintuitive behavior was simply something that Galileo observed. He didn't have a theory that really explained it. His observation implies that heavier objects resist changes in motion, but explaining this was not really possible until Newton came along.

NEWTON AND GRAVITY

- In 1666, at the age of 23, the future Sir Isaac Newton took the first quantifiable and scientific step in mankind's long quest to devise a theory of everything.

- Before Newton turned his attention to the problem, people didn't understand gravity in the way that we do now. Medieval scholars still thought that everyday gravity on Earth was essentially a property of matter. And this had nothing to do with the motion of the heavens, which was thought to be governed by a different power—a higher power.

- The connection of the stars with the religious idea of Heaven kept them quite apart from pedestrian and terrestrial things. In fact, the idea that earthly and celestial gravity were the same had never really crossed anyone's mind.

- But Copernicus had set the Sun at the center of the universe, an idea supported by Galileo's observation of the moons of Jupiter. Galileo also found that all objects fell at a certain uniform rate, regardless of how heavy they were. Johannes Kepler had quantified the period of the orbits of the planets as being related to their distance from the Sun.

- It was with these ideas that Newton turned his attention to gravity. He wondered if maybe the force that caused an apple to drop from a tree could perhaps reach out and tug at the Moon to guide it in its path around the Earth.

- Newton took the first step toward a unified theory when he grasped the connection between the existing theory of terrestrial gravity and the celestial gravity that governed the heavens. Two phenomena that had long seemed distinct were shown to have a common origin.

- His basic reasoning was the following. One can see an apple fall. Going a bit further, if you shoot a cannonball in a horizontal direction, it also falls, but travels a distance before it hits the Earth. Shoot it with more velocity and the cannonball goes even farther before hitting the ground.

- Taking this idea to its logical extreme, Newton reasoned that if you shoot the cannon with enough velocity, the cannonball will continue to fall toward the center of the Earth but will end up moving in a circular path until it orbits the planet and hits the cannoneer in the back.

- Newton's brilliant insight was to make the connection between the Moon and the cannonball. He reasoned that the Moon was constantly falling toward the Earth, but it had enough motion that it never landed. Instead, it was held in an eternal orbit. With that insight, the law of universal gravitation was born.

- Newton hypothesized that the force between 2 objects was proportional to the product of the mass of those 2 objects divided by the distance between them squared. If we use *F* to symbolize force, *m* to denote mass, and *r* to represent the separation between 2 objects, we can mathematically write Newton's equation as follows:

$$F \propto \frac{m_1 m_2}{r^2}$$

- His assumption could then be tested using his newly invented calculus, and he found that he could predict astronomical phenomena. His theory had been tested and confirmed in the most satisfying of ways—by data and observation.

- Newton said that gravity would be weaker the farther you were from the center of the planet. And we find that if a person's weight at sea level is 150 pounds, then his or her weight on the world's highest peak is only about half a pound lighter.

- Newton's theory is consistent with the conjecture that gravity is basically constant on the surface of the planet. Observing the variation predicted by Newton would have to wait until better instrumentation was developed.

- While Newton's first ruminations on the nature of gravity occurred in 1666, his theory was not published in a formal way until 2 decades later, when he published his *Principia Mathematica*. Working out the implications of a big idea and verifying that it is true can take a long time.

INERTIAL VERSUS GRAVITATIONAL MASS

- Newton also achieved another sort of unification and didn't even know it. This was a unification of 2 aspects of mass—what you might even call 2 kinds of mass—and it had huge consequences 250 years later.

- The fact that all objects fall at a single rate was demonstrated by Galileo. The idea has been proven again and again. Yet the fact that all objects fall at the same speed, irrespective of whether they are heavy or light, isn't at all obvious. But the fact that they do fall at the same speed holds a deep message.

- When an object falls, there are 2 crucial and distinct facets of the nature of mass. The first facet is that mass is the quantity that causes the force of gravity. If you increase an object's mass, it feels a greater weight. Weight is just a downward force. So, one way to think of mass, perhaps a modern way, is that it is like a charge of gravity.

- Think about electricity. In electricity, objects have an electric charge and the force they feel depends on the amount of that same electric charge. More charge means more force. So, in terms of gravity, mass is the gravitational charge. This form of mass is called gravitational mass.

- But mass has another property—specifically, inertia. Even in the absence of gravity, mass resists changes in motion. For example, imagine trying to push a Styrofoam ball floating in water and compare that to the effort it would take to push a battleship. Obviously, it's more difficult to push the battleship, and that's because it has more mass and therefore more inertia. This isn't merely an effect of gravitational charge because it's true even in deep space, where there is no appreciable gravitational force. This kind of mass is called inertial mass.

- Newton never imagined that there were 2 types of mass. Because he only imagined a single kind of mass, he treated the 2 as equal. This leads to the notion that the acceleration—the motion—of an object orbiting the Earth doesn't depend on the mass of the object.

- Given that we've measured and experimentally proven that the acceleration due to gravity doesn't depend on the mass of the object, we've actually said something very fundamental about the universe: that gravity and inertia are somehow linked. This is a sort of unification, but it wasn't explicitly faced until Einstein began his investigations into general relativity.

- Newton's theory of gravity explains Galileo's observation that all objects fall at the same rate—or, more accurately, Newton's theory of gravity makes clear that Galileo's observation that all objects fall at the same rate requires that inertial and gravitational mass be the same. It turns out that Newton's theory can also explain all of Kepler's laws.

- Newton's theory not only unified celestial and terrestrial gravity, but it also provided a strong mathematical underpinning for observations of the behavior of astronomical bodies that were made before his time. Newton had no idea what caused gravity—and he famously said that he would offer no "hypotheses"—but his theory of gravity's effects was spectacularly successful.

- Newton was the first person to formally and analytically demonstrate how terrestrial and astronomical gravity were the same thing. In so doing, his work became the first example of a big unification within physics, and of physics with math. Like Maxwell's theory of electromagnetism, it is a cornerstone example of what it means to unify previously separate phenomena with a single theory.

- The fact that he tacitly equated inertial mass and gravitational mass is actually of enormous importance, although the importance of that insight would not be recognized for 250 years, when Einstein unified everything Newton accomplished with his theory of special relativity to make possible our modern-day theory of gravity.

READINGS

George Gamow, *Gravity*.
Pask, *Magnificent Principia*.

QUESTIONS

1. The International Space Station orbits at an average height of about 250 miles above the Earth. What is the relative strength of gravity there as compared to the surface of the Earth?

2. If the number you determine is not 0, then why do astronauts appear to be weightless?

EINSTEIN'S GRAVITY BENDS SPACE-TIME

17

While there is no question that Sir Isaac Newton's insights were a huge step forward in our understanding of gravity, from a modern perspective, his ideas were incomplete. In fact, they were inaccurate in certain situations. For example, Newton's laws work very well when gravity is relatively weak. However, near the surface of stars, where gravity is much stronger than we experience on Earth, an improved theory is necessary. For this, it took another brilliant mind to advance our understanding: Albert Einstein.

BENDING SPACE AND TIME

- Einstein's contribution to our understanding of the nature of gravity was quite different from Newton's. Newton had created a unified theory of gravity by realizing that 2 seemingly very different phenomena were really the same thing. Einstein's insight was enormously deeper and more profound, in that he united all of gravity with the nature of space and time—things that had seemed quite separate from gravity.

- Our intuitive, everyday idea goes like this: Every day we live. We walk around from place to place and experience the moments as they tick by. From these experiences, we build our intuition of space and time. Space is the location where things occur, and time measures the duration of events.

- Einstein's first insight, when he developed his theory of special relativity in 1905, was that space and time were really 2 different components of a deeper and more fundamental construct, called space-time. Basically, in the same way you can think of north/south and east/west making up a map, space and time make up space-time.

- Einstein had previously used space-time to explain some of the mind-blowing things that happen when an object goes very fast, such as clocks slowing down and objects shrinking.

- Einstein realized that space is not static. Space itself could bend and distort.

- Einstein was interested in the nature of gravity. In Newtonian physics, gravity is caused by massive objects. Einstein revisited the idea of space-time that he developed with his teacher Hermann Minkoswki as a consequence of special relativity. Einstein realized that space-time was malleable and that mass could distort space-time.

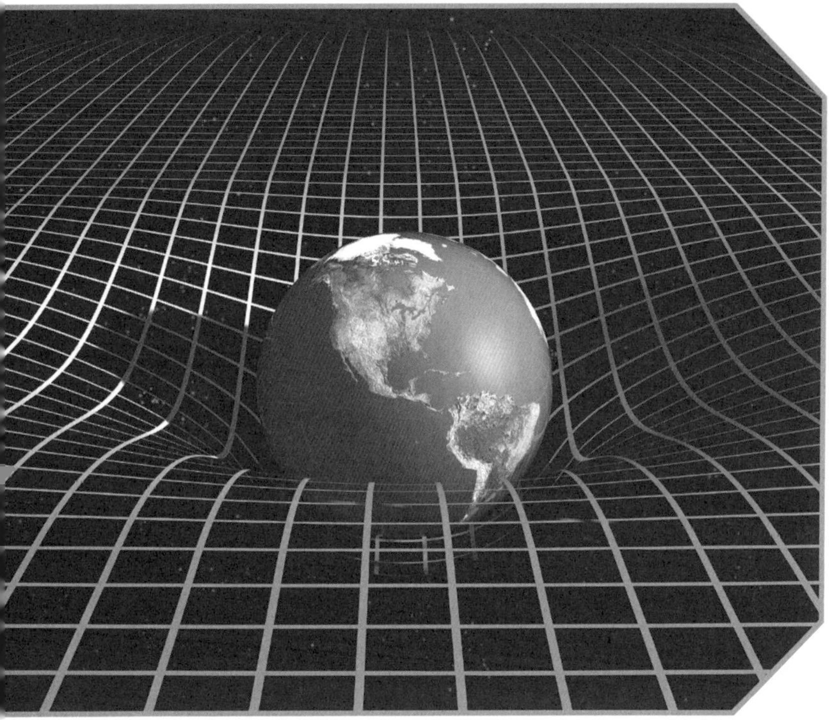

- A common way to show this is to imagine a large stretchy membrane with a grid on it. If you put a large mass in the center of the membrane, the membrane stretches downward, distorting space. This theory of gravity is called Einstein's theory of general relativity.

- In 1919, just 4 years after the theory was written down, Sir Arthur Eddington used a solar eclipse to test Einstein's theory. While the daytime Sun is so bright that we can't see the stars, we know that there are stars on the far side of the Sun. However, during an eclipse, the light from the Sun is blocked, and we can see those distant stars.

- During the 1919 eclipse, Eddington photographed the position of stars in the Hyades cluster, including Kappa Tauri in the constellation of Taurus. When the light from those distant stars passed near the Sun, their light traveled through space distorted by the Sun's mass. Because that light journeyed through bent space, the stars appeared to be in a different location than they would have been in if they had passed through space that wasn't bent.

- Modern astronomers believe that Eddington's measurements were probably not precise enough to support his enthusiastic claims on Einstein's behalf, but more modern measurements have confirmed Eddington's conclusions.

- Another test that verified general relativity is the finding that the orbit of Mercury isn't a simple ellipse. The point at which Mercury is closest to the Sun moves over time. This is called the precession of the orbit of Mercury, and it is caused mostly by the gravity of the more massive planets.

- If you calculate Mercury's orbit using Newton's theory of gravity, you find that it doesn't perfectly match reality. By contrast, Einstein's theory gets it right. And there are many other reasons that we believe in the theory of general relativity.

- In particular, we are interested in a unified theory of everything, and general relativity has many things to offer. First, general relativity explained gravity in a way that Newton left mysterious. Newton's theory imagined that some kind of force somehow reached out across space and affected a distant celestial body. That's a perfectly workable way to think about gravity for many purposes, but it is somehow unsatisfying. It suggests that things that seem unconnected actually are somehow connected.

- Einstein's improved theory of gravity fixes this issue. It's not that the 2 objects really interact with one another, and it's not that the 2 objects have no understandable connection. Rather, the 2 objects bend space. In the case of the Earth and the Moon, the more massive Earth does most of the bending of space, and the Moon travels in that bent space.

- Einstein's theory also unified how the location of matter and energy was tied to the very geometry of space itself. You can see that in a mathematical way by looking at this equation, where $G_{\mu\nu}$ describes the geometry of space and time, $T_{\mu\nu}$ describes the distribution of matter and energy, and the remaining term is just a constant.

$$G_{\mu\nu} = \frac{8\pi G}{c^4} T_{\mu\nu}$$

- This equation says that the geometry of the universe is just a constant times the mass and energy distribution in the universe. Thus, the shape of space and the distribution of matter within space are intimately connected. Stating the meaning of general relativity at its most simple level, matter and energy determine the shape of space and time.

- This idea has some very deep implications into the very fabric of reality. It means that any theory of everything we come up with will have to exist in a malleable universe, in which the objects inside the universe have an effect on the very warp and weave of the cosmos.

MODERN TESTING OF GENERAL RELATIVITY

- The problem with general relativity is that the predictions agree very well with old-fashioned Newtonian gravitational theory until the gravitational field becomes very strong or the instrumentation becomes very precise.

- If the gravitational field becomes very strong, the relativistic effects become very large, while if the instrumentation is very precise, this allows for the measurement of even small effects on Earth. Both approaches have been pursued, and both sets of measurements support the theory of general relativity.

- Measurements done on Earth are true experiments over which scientists can have complete control over the circumstances. One such important measurement revolves around a very surprising prediction of general relativity: Clocks in stronger gravitational fields run more slowly than ones in higher ones.

- The effect on Earth is very small. Because of how gravity depends on distance from the center of the Earth, the gravity on Mount Everest is lower than at sea level. Assuming that Mount Everest existed for the entire lifetime of the Earth, which is 4.5 billion years, a clock on the top of the mountain would be running 39 hours ahead of a similar clock at sea level. That's over a period of more than 4 billion years, which is a tiny effect.

- But improvements in instrumentation give the ability to measure small effects, and we can now measure this change in the rate at which clocks tick. A classic experiment performed in 1971 by Joseph Hafele and Richard Keating involved taking ultraprecise atomic clocks on commercial airliners. They flew around the world twice, first eastward and then westward. They compared the time duration experienced by the clocks in the airplane to clocks that were stationary on Earth.

- Because the clocks in the airliners spent a lot of the time at high altitude, it is expected that they ran a little faster than the ones left on Earth. Airliners not only move to high altitudes, but they also simply move. And moving clocks move more slowly than stationary ones. So, Hafele and Keating had to take into account effects from both special and general relativity.

- They found that when the traveling clocks were reunited with their stationary kin, the clocks disagreed. Furthermore, the clocks disagreed precisely as predicted by special and general relativity theory.

- James Chou used a single aluminum ion as a clock, which could measure time to a precision of 1 part in 10^{16}. A pair of clocks using this technology were synchronized, and one of them was lifted a foot off the ground. The 2 clocks began to run at different speeds. The lifted clock was then brought back to the same height, and the clocks ran in synch. The process was repeated with the second clock, and the same results were observed.

James Chin-wen Chou operates NIST's Second Quantum Logic Clock

- But we've moved far beyond merely testing general relativity, which has a crucial role in the Global Positioning System (GPS). This system works by comparing the arrival time of signals from an orbiting constellation of a little more than a few dozen satellites. GPS satellites are both moving very fast and in a weaker gravitational field, so both special and general relativity have to be taken into account. GPS is an ongoing confirmation of general relativity.

- General relativity has measurable, and sometimes important, effects in the weak gravitational field of Earth, but the effects are much bigger near strong gravitational fields.

- Einstein predicted, and Eddington (or subsequent researchers) verified, that light was affected by gravity. On the face of it, this is already a surprising prediction. According to Newton's theory of gravity, the gravitational force between 2 objects depends on the product of the mass of the 2 objects. Light, being massless, should not experience gravity. But it does, and that changes things.

- Consider an object with mass—for example, a tennis ball. If you are on the surface of the Earth and you toss the ball upward, the ball slows down and then falls back to Earth. Throw it up faster and it goes higher before falling down. This continues as a pattern. However, there is a minimum velocity beyond which the ball never falls back down. This speed is called escape velocity. If you shoot an object upward at escape velocity, it can escape the gravitational field of the Earth and, at least in principle, wander the cosmos.

- In 1796, French scientist Pierre-Simon Laplace imagined that there might be objects, such as planets or stars, that were so massive that the velocity needed to escape them would exceed the speed of light. He imagined objects so massive that not even light could escape them. These objects are now called black holes.

- Laplace was wrong in detail. There are 2 ways you could get an object from which light couldn't escape: an object that is very large or a smaller-sized object that is very dense. In reality, black holes are objects with a few times the mass of our Sun, packed into tiny volumes.

- While black holes were predicted by Einstein in 1916, and their most basic mathematics were calculated by Karl Schwarzschild in the same year, the term "black hole" was coined in 1967 by John Wheeler. And the first one was only discovered in 1971. It's called Cygnus X-1, and it is the first X-ray source found in the constellation Cygnus. We can't see the black hole itself; after all, light can't escape from it. But black holes can affect the motion of nearby stars.

- Black holes are fascinating, but for purposes of a theory of everything, they are simply a demonstration that Einstein's theory of general relativity is a good unifying theory, even for gravity at its most intense.

- There is one more prediction of general relativity that was definitively demonstrated in 2016, a full century after black holes were predicted.

- Einstein predicted in 1916 that if you had 2 very small and very massive astronomical bodies orbiting each other at very small distances, the constant motion of these 2 masses would send gravitational waves across the cosmos. Basically, these waves would temporarily shrink and expand space, making distances longer and shorter.

- Studies in 1974 of a pair of ultracompact stars called neutron stars suggested that they were emitting gravitational waves as they orbited each other. This led to the 1993 Nobel Prize in Physics. But this observation was indirect.

- In 2002, Laser Interferometer Gravitational-Wave Observatory (LIGO) began operations to look for gravitational waves. On September 14, 2015, the scientists working on LIGO observed what looked very much like a gravitational wave event. This meant that a long time ago and in a faraway galaxy, 2 black holes orbited each other furiously, losing energy to gravitational radiation, and they eventually merged.

READINGS

Carroll, *Spacetime and Geometry.*
Collier, *A Most Incomprehensible Thing.*
Einstein, *Relativity*, part II.
Gamow, *Mr. Tompkins in Paperback*, chaps. 4–5.

QUESTIONS

1 Does the recent discovery of gravitational waves have any consequences on the quantum properties of gravity? Why, or why not?

2 If the Sun were to suddenly turn into a black hole (it can't), what would be the consequences on Earth? Would the Earth fall into the ravenous clutches of the black hole? If not, why not?

18 WHAT HOLDS EACH GALAXY TOGETHER: DARK MATTER

An all-encompassing theory of everything must explain all phenomena—from quarks to the cosmos. Dark matter is an enormous mystery. It's a huge component of the energy and matter content of the universe; it dominates more than our familiar form of matter. This means that we can't possibly hope of inventing a theory of everything that doesn't solve this particular conundrum.

COSMIC PINWHEELS

- In 1929, American astronomer Edwin Hubble was able to definitively state that the large whirlpool seen in the constellation Andromeda was actually a separate galaxy. With that observation, our understanding of the cosmos changed. Individual galaxies, islands defined by gravity in the sea of the cosmos, were found everywhere.

- This observation also changed our understanding of our own Milky Way galaxy. The realization that the Milky Way was just a large and gravitationally bound group of stars led scientists to apply the known laws of physics and the observed amount of matter to understand our galactic home better.

- By 1932, Dutch astronomer Jan Oort had measured the rotation of the Milky Way and found that it was rotating faster than could be explained. It looked like the galaxy was half as massive as it needed to be for the observations and theories to hang together.

- The following year, Swiss astronomer Fritz Zwicky was studying a large cluster of about 1000 galaxies, called the Coma cluster. By studying the motion of the galaxies and calculating the total mass contained in the cluster, he determined that to keep all those galaxies gravitationally held together in a single cluster, there needed to be about 400 times more mass than could be seen.

- Modern measurements say that the invisible matter is lower than Zwicky claimed, but even now astronomers agree that the cluster contains more mass than can be directly observed. Zwicky invented the term "dark matter" to describe whatever it was that couldn't be seen.

- While Zwicky's measurement was the first unambiguous observation of a phenomenon that we continue to explain by dark matter, perhaps the clearest evidence was reported in the 1970s by American astronomers Vera Rubin and Kenneth Ford, who looked at a large number of galaxies and measured how fast they rotate.

- Astronomers can approximate the distribution of mass in galaxies, and when they do a careful job of working out the mass distribution, they can make predictions of the orbital speed of stars at different orbital radii.

- The orbital velocity is zero at the center of the galaxy, gradually rises until some maximum near the edge of the galaxy where the stars peter out, and then drops off and approaches zero when one looks at straggler stars and clouds of hydrogen gas that are far beyond any clear outer edge of the galaxy.

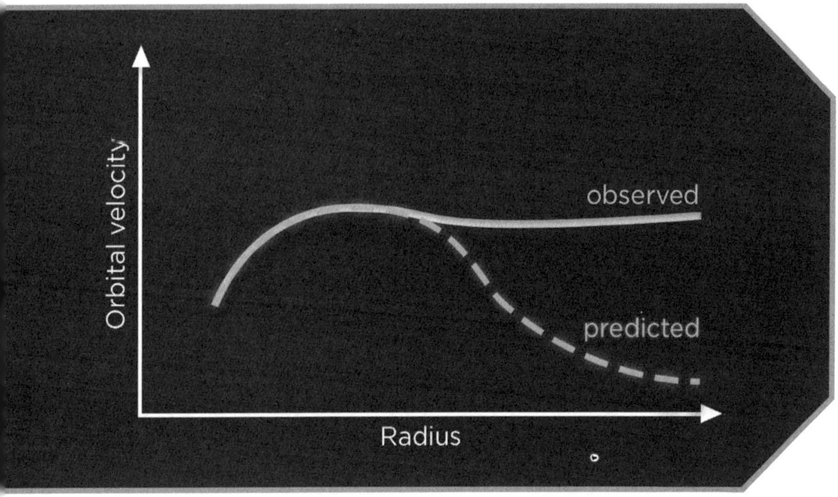

- If we look at stars near the center of the galaxy and plot their measured orbital velocity, the measurement and prediction agree very well. Even as we approach the middle of the galaxy, the degree of agreement between data and theory is satisfactory.

- However, as we get to the outskirts, where most of the visible stars have petered out, there is a problem: The orbital velocity of stars and hydrogen gas clouds is more or less constant, independent of the orbital radius. This is in stark contradiction with the prediction.

- Nearly all galaxies have stars far from the center that are orbiting much too quickly. If the only force that the stars were feeling was the force of gravity due to visible matter, then gravity couldn't hold onto them. Galaxies would tear themselves apart.

- This showcases a horrible disagreement between data and theory, and that means that our theory of gravity, even general relativity, is missing something in the world of gravitational attractions.

- The force causing stars to move in a circular path is gravity. This means one of 3 things: We don't know what makes objects move in a circle, we don't understand gravity, or the forces aren't equal.

- On the side of kinematics, a proposal in 1983 by Israeli physicist Mordehai Milgrom suggested that in the outskirts of the galaxy, where the force due to gravity is really small, Newton's law $F = ma$ doesn't apply. He picked a form that was chosen to solve the discrepancy of the problem of orbital speeds of galaxies. The term for this approach is called modified Newtonian dynamics (MOND).

- On the gravity side, there were 2 possibilities. One is that the laws of gravity were wrong. Both Newtonian gravity and the theory of general relativity give identical results, and other modifications of gravity have been proposed, but none has satisfactorily solved all problems, so this approach is not considered viable.

- On the other hand, the idea of invisible matter seems promising. That was Zwicky's answer in the 1930s. And there are 2 possibilities. One is that the invisible matter was an unexpected amount of ordinary matter that was just dark, such as black holes and big planets. Because these objects were massive, compact, and existed in the halo, or the periphery of galaxies, they were called massive astrophysical compact halo objects (MACHOs).

- However, experiments that began in the early 1990s ruled out that idea. Basically, they looked at stars in the Greater and Lesser Magellanic Clouds and also at the center of the Milky Way. They were looking for instances when MACHOs passed in front of the stars. While instances were observed, not enough were observed to explain the dark matter problem.

- The other possibility was that maybe there existed some other kind of matter, one that hasn't been observed yet. This kind of matter would interact very weakly but still have mass. These hypothetical objects were called weakly interacting massive particles (WIMPs).

- WIMPs are imagined to be subatomic particles. To be dark matter, they would have to have mass. They would have to be stable and not decay for tens of billions of years, otherwise they would have decayed away. They need to be electrically neutral, otherwise they would have interacted with light from distant stars on its way to Earth. They can't contain quarks, otherwise cosmic rays in space would hit them, and we would see emissions of radiation.

- After nearly a century, we were left with 2 viable possibilities: MOND and the existence of WIMPs.

- Dark matter was invoked by Fritz Zwicky to explain the fact that the Coma cluster didn't fly apart. That means that dark matter is found throughout clusters of galaxies. The other things found in clusters of galaxies are the stars and galaxies, but also hydrogen gas floating around in space in between the stars and in between the galaxies.

- The question of what would happen under the MOND and dark matter hypotheses if 2 clusters of galaxies collided was answered in 2004, when the Bullet Cluster of galaxies was found. The Bullet Cluster is actually 2 clusters of galaxies that passed through one another a long time ago.

- Using ordinary telescopes, the location and mass of the galaxies were identified. Observations using X-rays between the 2 galaxies showed that the diffuse hydrogen gas in the 2 clusters collided like 2 gusts of wind. When the clouds hit, they heated each other up, and that's why they could be seen using X-rays.

- If dark matter were somehow associated with the hydrogen gas, we'd expect to see it located at the center of the colliding clusters. That's because the hydrogen gas has more mass in it than the stars in the galaxies.

- If what we are calling dark matter is associated with the hydrogen gas, that would be consistent with the MOND hypothesis. On the other hand, if the dark matter did not interact like hydrogen, then we'd expect to find that the dark matter was located at the center of the 2 clusters of galaxies.

- Using even more distant galaxies and studying how their images were distorted as the light passed through the Bullet Cluster allowed astronomers to find where the dark matter mass was found. This technique is called gravitational lensing because gravity bends light almost like a lens. It is often used in situations where we have a massive foreground object of interest and a lot of light-emitting stars or galaxies in the background.

- When the observations were performed, it was found that the dark matter didn't participate in the hydrogen gas collision but was rather found located at the center of each of the 2 colliding clusters of galaxies. This was a ringing endorsement for dark matter and a significant setback for the MOND proponents.

- We know that dark matter seems to exist in particle form, as opposed to more adventurous and speculative ideas that include matter that isn't quantized. The particles are electrically neutral, emit no electromagnetic waves, contain no quarks, and are stable over the lifetime of the universe. They interact gravitationally, but definitely not with the electromagnetic and strong nuclear forces. For many years, there remained the possibility that dark matter could interact via the weak nuclear force, but this was pretty much ruled out in the years after 2010.

- When we study the universe as a whole, we can determine the amount of dark matter that exists. There is about 5 times as much dark matter as there is ordinary matter, which includes the stars and galaxies but also the huge amount of hydrogen that isn't bound up in objects that glow. Of the ordinary matter, only about 10% is in stars and galaxies; the remaining 90% is hydrogen gas.

THEORETICAL SPECULATION

- The simplest idea of dark matter is that there exists a single kind of particle just floating around in space. The mass of the particle isn't known, but it's maybe 10 or maybe 100 times heavier than a neutron—which does not equate to a whole lot of mass. So how can there be so much dark matter?

- About a fifth of all ordinary matter is densely concentrated in planets and stars. Dark matter is more like unbound hydrogen: A lot of it is found in the empty space between the planets and in interstellar space. Even broader, dark matter is found in huge spherical clouds around and beyond each galaxy, with the clouds extending 10 times farther than the stars we can see. The density might be very small, but the volume is huge, and that's what makes the cumulative mass so big.

- There are many ideas about what dark matter might be, but the principle of supersymmetry has been perhaps the leading candidate. Supersymmetric theories predict that for every known particle of the standard model, there is a supersymmetric cousin. Furthermore, because we haven't found supersymmetric particles, we know that they must be much heavier than their standard model counterparts.

- Given that photons, Z bosons, gluons, and Higgs bosons are all electrically neutral, their supersymmetric cousins must also be neutral. And none of those particles has quarks, so supersymmetry has been looking promising.

- Different supersymmetric models will predict that different supersymmetric cousins are heavier and others are lighter. However, all models that conserve R-parity (a quantum number that is +1 for ordinary matter and −1 for supersymmetric particles) have a lightest supersymmetric particle, which must be electrically neutral or we would have seen it already. Thus, if supersymmetric particles exist, then the lightest supersymmetric particle could well be dark matter.

- Supersymmetry was invented as the last possible space-time symmetry that a theory could have. If it is correct, it explains why the mass of the Higgs boson is so light. And it could explain the dark matter problem. But we have to find evidence that the explanations are real before we can get too excited.

EXPERIMENTAL APPROACHES

- There are 3 distinct ways that scientists are looking for dark matter. One is an attempt to observe dark matter as it passes through the Earth. This is called direct detection. This involves putting detectors in deep mines, half a mile or even a mile under the surface.

- A second method is to look for places where dark matter is denser than on Earth. An obvious such place is the center of the Milky Way. If dark matter is located there, then maybe dark matter particles will interact and cause ordinary particles, such as photons, to be emitted. We could look for high-energy photons coming from the center of the galaxy to appear in quantities greater than we expect. This approach is called indirect detection.

- The third method of finding dark matter is to make it from energy. This is what we do in particle accelerators like the Large Hadron Collider. We smash high-energy particles together and hope that the energy is converted into dark matter particles. This is the same way we found the Higgs boson and many other particles. As long as dark matter interacts with ordinary matter with a force that is stronger than gravity, it should work. This approach is called experimental detection.

- There is an entire cottage industry of scientists using all 3 approaches to find dark matter. There have been many false sightings that disappeared under the scrutiny of more research. In addition, it will take a combined set of observations from all 3 techniques to convince scientists that dark matter has been found.

READINGS

Gates, *Einstein's Telescope*, chaps. 1–9.
Hooper, *Dark Cosmos*, chaps. 1–6.
Nicolson, *Dark Side of the Universe.*
Panek, *The 4 Percent Universe.*
Randall, *Dark Matter and the Dinosaurs.*

QUESTIONS

1 Which of the possible explanations for the observations explained by dark matter do you find plausible?

2 If dark matter exists in the solar system (as it is thought to), why isn't it obvious and easy to find?

19 WHAT PUSHES THE UNIVERSE APART: DARK ENERGY

Beyond dark matter, there's a second cosmic conundrum that dwarfs both dark and ordinary matter combined. There is a third substance, called dark energy, which seems to be a ghostly energy field that will determine the very fate of the cosmos. There is no credible likelihood that we will ever create a theory of everything until we get a handle on the dark cosmos.

BREAKNECK EXPANSION

- The big bang idea says that the universe began 13.7 billion years ago, when the universe was much more compact. Points in space that are now separated by a light-year, for example, were once much closer to one another—indeed, closer than adjacent protons in an atomic nucleus.

- We know from looking at distant galaxies that they are moving away from us, and from that, we can infer that the universe is expanding. And if we ran the film backward, we imagine that the entire visible universe was once concentrated into a tiny volume—maybe approaching a mathematical point with zero size. Something pulled the trigger, and the expansion started. It was an explosion of space-time.

- In the beginning, an explosion—for example, of a firecracker—is very hot, and it expands and cools as time goes on. And in the case of the universe, gravity comes into play. Gravity is an attractive force. So, in the case of the big bang, we expect that after the initial explosion, the universe will be expanding at high speed. However, as time goes on, gravity will slow down the expansion.

- We can combine bits of information—the recessional velocity of a galaxy, its distance, and, from that, how far back in time it is—and use the data to figure out how the expansion of the universe unfolded. It seemed like there are only 3 possible scenarios.
  1. The expansion of the universe slows down but never stops; it just goes on forever and never becomes zero.

  2. The expansion slows down so that as time becomes infinite, the expansion just stops.

  3. There is so much gravity contributed by the matter of the universe that the expansion of the universe not only slows down to zero in finite time, but it then begins to shrink, as the force due to gravity finally pulls it all back together. Astronomers call this the big crunch.

- In the late 1990s, 2 groups were trying to find out which of these possibilities described the actual universe, and in 1998, they announced the answer. It was a fourth scenario: The expansion of the universe was accelerating. There is no way you can explain this with our intuitive understanding of gravity.

- In the ensuing years, researchers went back to the beginning of the universe and built a timeline showing how the rate of expansion has changed over time so far. It turns out that for the first 8 or 9 billion years, the expansion occurred pretty much as expected: first fast, then slower. However, about 5 or 6 billion years ago, the expansion started to speed up. Something happened that overpowered the expected effects of gravity.

EINSTEIN'S BIGGEST BLUNDER

- When Einstein invented general relativity in 1915, the opinion of the day was that the universe was eternal and had existed forever. And that was a problem for Einstein.

- He had a robust theory of gravity worked out. He knew that gravity was an attractive force. If you start with a uniform distribution of matter as a starting point for his equations of general relativity, gravity would cause the universe to collapse in on itself. Basically, he knew that his equations said that a big crunch was inevitable.

- To save his theory, he needed some sort of repulsive force to balance gravity and cause the universe to be static and eternal. Recall Einstein's equation:

$$G_{\mu\nu} = \frac{8\pi G}{c^4} T_{\mu\nu}$$

- The left side describes the geometry of space-time. In contrast, the right side has $T_{\mu\nu}$, which describes the distribution of energy and matter throughout space, and the remaining term is just a constant. Basically, it said that geometry equaled the distribution of energy.

- To save general relativity, in 1917, Einstein made a small modification to his equation. He added a single term to the left side of the equation:

$$G_{\mu\nu} + \Lambda g_{\mu\nu} = \frac{8\pi G}{c^4} T_{\mu\nu}$$

- The $g_{\mu\nu}$ is called the metric tensor, which is basically the baseline geometry without energy distortions. It describes space-time without taking into account the distortions caused by energy and mass. The metric tensor is multiplied by Λ, which is a constant, or a number, called the cosmological constant.

- The cosmological constant is a constant energy density located throughout space-time. It means that Λ is a uniform form of energy throughout the universe, but it is carefully chosen to provide a repulsive form of gravity. It was invented to counteract the repulsive aspect of ordinary gravity and to make the universe static and unchanging.

- In 1927, Georges Lemaître suggested—and in 1929, Edwin Hubble confirmed—that the universe was not static. It was expanding. And with the changing universe, the philosophical need for the cosmological constant went away. Einstein removed the cosmological constant term and supposedly called it his biggest blunder.

- Einstein publicly removed the cosmological constant in about 1930, and for more than 60 years, scientists largely forgot about it. That all changed in 1998, with the discovery that the expansion of the universe was accelerating. It appeared that there was a repulsive form of gravity that needed to be taken into account. And the cosmological constant made a comeback.

- The cosmological constant isn't generally accepted yet. The reason is that the cosmological constant is a very specific model. It says that there is a constant and uniform energy density throughout the universe. There are other options. There might be a new form of energy throughout the universe that is not a constant.

- But if this new energy might not exactly be the cosmological constant, then it needs a name. American astronomer Michael Turner coined the expression "dark energy" before April 1998 as a generic moniker for this mysterious, universe-spanning energy field.

- Unfortunately, the terms "dark matter" and "dark energy" sound a lot alike, and this is confusing to many people. Dark matter and dark energy are actually very different things, where what they do have in common (so far, at least) is that we don't know much about either of them.

- Dark energy is about 70% of the mass-energy of the universe. Ordinary matter is 5%, and dark matter makes up the remaining 25%. Dark energy is the largest energy component of the universe. And we know very little about it.

- Ordinary matter and dark matter are just matter. There is a fixed amount of matter in the universe. But as time goes on, due to the expansion of space, the mass density will decrease.

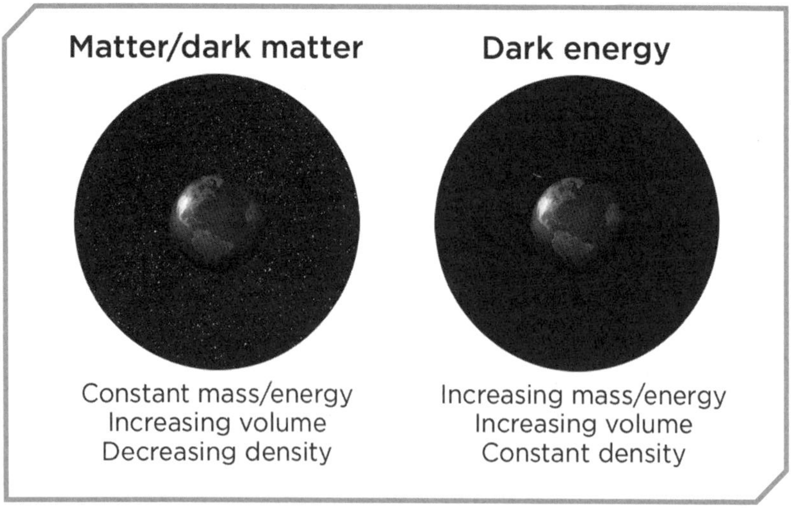

- Dark energy is a constant density, not constant energy. Thus, as the volume of space increases, the density of dark energy remains constant. Because the volume of space gets bigger, the quantity of dark energy increases.

- This seems to violate the law of conservation of energy, but in general relativity, even ordinary energy is not necessarily conserved.

- The fact that dark energy has constant density means that early in the universe, when the volume was small, the amount of dark energy was also small. And because ordinary and dark matter are constant, during the early times in the universe, the amount of dark energy was so small that is was essentially irrelevant. However, as the universe began to expand, dark energy became an increasingly large portion of the universe's energy budget. And dark energy will become more and more dominant as time goes on.

- This dark energy question brings in a huge issue regarding an ultimate theory. The problem arises when you try to imagine a source for dark energy and look for a way to tie this energy to what we know about empty space when we talk about quantum fields.

- One possible resolution leads us back through quantum electrodynamics (QED), one of the most thoroughly tested theories, which predicts the interactions of charged particles and photons.

- One consequence of QED is a prediction that matter and antimatter particles are constantly being created and destroyed. This idea sounds counterintuitive, but it is only by including the existence of these particles in QED calculations that scientists can achieve such incredible agreement between prediction and measurements. The frenzied activity of these ephemeral particles can be expressed as an energy, and the name of this energy is vacuum energy, or zero-point energy.

- Given that scientists know that vacuum energy is real, we are able to think about tying together the quantum and cosmic realms. The connecting idea is simple: Dark energy is an energy field that spans all of space. This is also true of vacuum energy. Therefore, it stands to reason that maybe vacuum energy is the source of dark energy. But the energy of the ephemeral particles can be calculated, and it is huge. Vacuum energy is way too big.

- This is an extraordinary mystery. QED works so incredibly well that we know that vacuum energy exists. But we also know that dark energy is ultratiny. The 2 just don't agree. The name for this is the vacuum catastrophe.

- Any time you get something so outrageously wrong, you know that there is a deep and fundamental problem somewhere.

- The prediction of the energy density of vacuum energy depends on some assumptions that make sense but have never been proven.

- The vacuum catastrophe disagreement is just saying that we shouldn't be content with the standard model and general relativity.

THE NONCONSTANT CONSTANT

- While there is nearly universal consensus that dark energy exists in some form or another, there is considerable latitude for exactly what nature it takes. The cosmological constant is the easiest hypothesis, but it isn't the only one.

- The cosmological constant is a constant. It is an energy density that doesn't change in time and doesn't change in space. But what if the constant wasn't actually constant?

- The modern name for this possible phenomenon is called quintessence, a name that was proposed in 1998 in a paper written by Robert Caldwell, Rahul Dave, and Paul Steinhardt. Quintessence is an energy field that can change over time. We know that it can't change very quickly because when we compare how far away distant galaxies are and how fast they are moving, dark energy has to be pretty constant. But it doesn't have to be exactly constant.

- This distinction—perfectly constant versus nearly constant—has potentially big conceptual implications for how to figure out a theory of everything. If it is constant, then perhaps it is nothing more than a property of space—kind of a definition of space, meaning that if you have space, you have energy.

- But if it is changing with time, then maybe that means something else. Maybe it means that dark energy isn't a necessary consequence of space existing. Maybe dark energy has some other cause and that higher-level cause creates both space and dark energy.

- This question of whether dark energy is strictly constant or just changing slowly has profound consequences for predictions about the past and future of the universe. For example, if dark energy follows a cosmological constant, then the expansion of the universe will continue to accelerate. That means that very distant galaxies will move faster and faster away from us, until the space in which they sit is moving away from us faster than the speed of light. When that happens, they will disappear and we will never see them again.

- However, if the quintessence idea is true and dark energy is increasing, then the expansion will be an even stronger effect. Not only will distant galaxies disappear, but the expansion will overcome the gravitational binding of the local galaxies and maybe even pull apart the Milky Way, the solar system, Earth, and even atoms.

- This phenomenon is called the big rip, and it means that in the very distant future, all of matter will be so separated that individual atoms will float around with the distances between them and their neighbors so big that no other atoms can be seen.

READINGS

Gates, *Einstein's Telescope*, chaps. 10–12.
Hooper, *Dark Cosmos*, chaps. 7–10.
Nicolson, *Dark Side of the Universe.*
Panek, *The 4 Percent Universe.*

QUESTIONS

1 If dark energy changes with time and is increasing, what are the consequences for the future of the universe?

2 If the universe is expanding, why isn't the solar system expanding?

QUANTUM GRAVITY: EINSTEIN, STRINGS, AND LOOPS

20

How will we unite the known large-scale properties of the universe with the microscale successes of the standard model? In other words, how will we merge gravity and quantum mechanics? We simply don't know. But in our journey to devise a theory that explains everything, we must be able to explain gravity at small-distance scales. In this lecture, you will discover what we know about gravity and why things seem to fall apart at small distances. You will also learn about some ideas that look promising as we try to pull the force that governs the cosmos into the theoretical framework that supports all other understood phenomena.

GRAVITY THE WEAKLING

- A theory of quantum gravity doesn't already exist because the force due to gravity is outrageously weak. Gravity is so outrageously weak compared to the electric force that we will never be able to measure it in an atom. In fact, it's only because of the fact that the Sun and the Earth contain huge numbers of atoms—and because both of them are basically electrically neutral—that gravity is what guides the planets' orbit and not electromagnetism.

- Because gravity is so weak, it is enormously difficult to imagine how we can do experiments that will help us understand its quantum nature. Why do we think that gravity even has a quantum nature?

- All of the other forces are quantized, so it would seem weird if gravity didn't.

- The next reason comes from the conundrum of black holes, which are the remnants of very heavy dead stars. There is a minimum mass for a black hole formed in this way—2 or 3 times that of our Sun. And according to the theory of general relativity, all of that mass is concentrated in a single mathematical point with zero radius.

- As you get closer and closer to a black hole, gravity becomes stronger and stronger. In that sense, black holes aren't different from stars. However, at a certain distance from the center of the black hole, the strength of gravity becomes so high that not even light can escape. That distance is called the Schwarzschild radius. It depends on the mass of the black hole but is typically a few miles.

- This radius isn't the radius of the black hole; it's simply the radius at which the black hole's gravity becomes strong enough that light can't escape. The black hole itself is supposed to have zero size.

- This is an absurd statement. If it were true, you'd have the mass of several stars concentrated in zero volume. This means that you would have infinite density. And that is absurd.

- If one has a physics theory that predicts infinities like this, you know that the theory has been pushed beyond its limits. Before you actually get to zero size, there has to be some other physics that comes into play and changes the rules.

- It is generally known among professional scientists that some new physics must come into play near the center of a black hole. But why does that physics eventually have to become quantized?

- According to classical electromagnetic theory, an electron orbiting a proton is being constantly accelerated, which means that the electron would emit electromagnetic radiation and lose energy. The electron would then spiral down to the proton in a split second. However, we know this doesn't happen, and quantum mechanics was invented to rescue the atom from a realm where the older theory no longer applied.

- A similar thing occurs if you think about the gravitational consequences of an electron orbiting a proton. While the effect is inconceivably small, the electron should also emit gravitational radiation. In direct analogy, there must be some kind of new physics that keeps the electron stably located around the center of the atom rather than emitting gravitational radiation. That physics is probably quantized, and it is quantized for the same reasons that went into the development of quantum mechanics.

- This doesn't mean that a theory of quantum gravity has to have the same character as quantum mechanics, but it does suggest that a new kind of quantum physics must come into play.

QUANTUM GRAVITY

- There is a ton that we don't know about a theory of quantum gravity and how it would get drawn into the mathematical framework of the other known forces. But it's not like we know nothing. To begin with, we know a great deal about the hypothetical quantum particle that is responsible for carrying the gravitational force. It's called the graviton, but it hasn't been discovered yet.

- In all of the forces that have been successfully quantized, meaning the strong and weak nuclear forces and electromagnetism, the theory predicts a quantum of force. Their known force quanta are the gluon, the W and Z bosons, and the photon, respectively. These particles are all bosons with spin 1. The photon and gluon are massless; the others are massive. Some have electrical charge, but the gluon and Z boson do not.

- If it exists, the graviton is a massless, electrically neutral particle with a spin of 2. In fact, if we ever find any massless particle with spin 2, we've found the graviton; no other particles can exist with those properties and not be the graviton.

- Without a specific theory of quantum gravity and even without knowing if quantum gravity is even real, we know the properties of the graviton.

- We also know more about what a theory of quantum gravity must be. Actually, we know what it isn't. We know that it can't simply blend ordinary quantum mechanics and general relativity. One reason is deeply embedded in the mathematics.

- After all, general relativity uses differential equations, which require that the functions involved must be smooth and continuous. Quantum mechanics, in contrast, is quantized and therefore discontinuous. Particles come in discrete lumps. This is the opposite of smooth and continuous. Mathematically, quantum mechanics and general relativity are incompatible.

- That turns out to not be an insurmountable problem. After all, this problem was kind of solved in quantum mechanics with ideas like wave-particle duality, in which subatomic objects move like waves but are detected like particles. Theoretical physicists could solve that particular problem.

- However, when physicists try to write a quantized version of general relativity, a problem arises. The quantized theory makes predictions that include infinities. Infinities arose when quantum electrodynamics

(QED) was being developed as well, when the simplest calculations said that the energy of an electron reacting with its own electric field was infinite. But it turned out to be pretty easy tame the infinities of QED. The trick was to hide that energy in the mass of the electron, which works because the mass of an electron is unchanging.

- That isn't true for the simpler ideas of quantum gravity. No neat mathematical tricks have arisen that somehow keep the beauty of general relativity and incorporate quantum mechanics. The 2 ideas just seem to be mutually incompatible. A theory of quantum gravity probably is not impossible, but it is devilishly difficult.

SUPERSTRINGS AND LOOP QUANTUM GRAVITY

- Admittedly, we don't really know what we're doing, but that doesn't mean that we haven't worked on the problem and that we haven't made some progress. There are 2 candidate approaches to quantum gravity that are both highly speculative, but centuries worth of accumulated thought have been put into them. These 2 ideas are called superstrings and loop quantum gravity.

- Superstrings started out in 1968, when Italian physicist Gabriele Veneziano was trying to understand the nature of the strong force. This was shortly after the quark hypothesis idea was proposed and before quantum chromodynamics was formulated and even before quarks were shown to be real entities. He noticed that a mathematical function called the Euler beta function could describe the strong force.

- Strings came into the picture in 1970, when Yoichiro Nambu, Holger Nielsen, and Leonard Susskind noticed that if they took some equations they were playing around with at the time and replaced a point-like particle with a small extended object—now called a string—then their equations were solved by the Euler beta function.

- These strings acted like small and stiff objects that could vibrate. The idea was that each type of vibration might correspond to a different kind of particle.

- The idea invoking strings was initially intended to be only a theory of the strong force—and it initially only incorporated bosons. In 1971, a young postdoctoral theorist named Pierre Ramond, along with others, was able to generalize the theory by incorporating fermions. He found that bosons and fermions came into the theory in pairs—a core tenet of supersymmetry.

- At that point, supersymmetric string theory was born, although it took a few years to demonstrate that Ramond's pairing was indeed supersymmetry. When that was achieved, the name "superstrings" became familiar to the theoretical community.

- A seminal moment occurred in 1974, when John Schwarz and Joel Scherk found a solution in these equations that predicted a massless particle with spin 2. Because any particle with those properties has to be a graviton, this got everyone's attention.

- It was the first time a quantum theory was able to incorporate a graviton without too much trouble. The problem was that there were all kinds of mathematical inconsistencies in the theory, and it took Schwarz a decade to work them out. In 1984, Schwarz and Michael Greene finally worked out those problems.

- A puzzling facet of superstring theory arose when people started trying to make real calculations of the theory. For decades, people had been saying that superstring theory doesn't make firm predictions, but some people puttering with the mathematics found that if they increased the number of physical dimensions in which the mathematics of the theory was cast, the mathematical difficulties became better behaved.

- They added more and more dimensions to the theory and found that when they got to 10 (3 spatial dimensions, 1 of time, and 6 necessary to make superstring theory work), the theory made physical predictions.

- By 1984, superstring theory had become a potential theory of everything. The core of the theory is that at the smallest and most fundamental scales, the ultimate building blocks are not mathematical points, as quarks and leptons are thought to be. Instead, they are small and vibrating objects. These objects are 1-dimensional, so they are shaped like little sticks of spaghetti or little hula hoops. The subatomic sticks of spaghetti are called open strings, and the hula hoops are called closed strings, and both are possible.

- The way the theory works is pretty simple: The known subatomic particles are not different objects but are simply different vibrations of a single and fundamental string. All particles can be imagined to be different vibrations.

- The vibrations occur in all of the 6 dimensions. We experience 4 dimensions (3 spatial and 1 time), so this dimension idea is really weird. There has to be something very different about these other dimensions or we can rule out the idea of superstrings. Those other 6 dimensions are thought to be very tiny—too tiny to see.

- Another proposed theory of quantum gravity is called loop quantum gravity, and it focuses less on the idea of particles and more on the idea of how general relativity explains gravity as curvatures of space and time. This suggests that a quantum theory of gravity could also be a quantum theory of space and time.

- While the implementation of this particular theory is very technical, it has a clear prediction, which is that at the very smallest sizes, space and time are themselves quantized. This means that there is a smallest-possible length and a smallest-possible time.

- One of the consequences of loop quantum gravity is that it predicts that the speed of light depends on the wavelength of light. Higher energy and shorter wavelength of light will travel more slowly through space than its longer-wavelength counterpart. The difference would be small, and studies are ongoing. So far, there has been no evidence that substantiates the idea of differing speeds of light for different wavelengths, but the idea remains possible.

- Neither of these theories is a well-developed candidate for a theory of everything. One theory focuses on space and time and ignores the other forces, while the other theory assumes space and time but brings the forces together. This means that each of these theories might be merely a piece of the final theory. Or both of them could be a piece—or neither.

READINGS

Green, *The Elegant Universe.*
Smolin, *The Trouble with Physics.*
Smolin, *Three Roads to Quantum Gravity.*
Susskind, *The Cosmic Landscape.*
Woit, *Not Even Wrong.*

QUESTIONS

1. What is the key barrier to making measurements of gravity at a particle accelerator like the Large Hadron Collider?

2. If space and time were quantized in the manner predicted by loop quantum gravity, what would the universe look like if you could shrink yourself down to a size scale 100 times larger than the smallest scale? What about to only 10 times larger than the smallest scale?

21 FROM WEAK GRAVITY TO EXTRA DIMENSIONS

The fact that the known forces change their strength as a function of energy and seem to have the same strength at specific energies is a strong indicator that the forces might eventually unify and a single force explains everything. Even if all the forces can be unified at an extremely high energy, our everyday question remains: Why is gravity so much weaker than all of the other known forces? This lecture will discuss the possibility that perhaps gravity is actually quite strong and the apparent weakness is simply an illusion due to an overlooked aspect of reality.

WHY IS GRAVITY WEAK?

- We know that we live in a world with 3 spatial dimensions, specifically left-right, up-down, and forward-backward. But what if gravity wasn't limited to 3 dimensions? Electromagnetism and the weak and strong nuclear forces would be stuck in 3 dimensions, but gravity might access 4 or 5—or more.

- Following that hypothesis, the basic idea is that gravity is essentially as strong as the other known forces and it only appears to be weak because it can and does spread out into additional dimensions. Under this proposition, gravity simply has more places to go.

- This is a mind-bending idea. How could gravity possibly have access to more dimensions than the other forces? It would certainly require new physics.

- What can data and measurements and what we know about geometry say about this particular supposition? It turns out that geometry tells us quite a lot.

- Newton's law of universal gravity states that the strength of gravity between 2 essentially point-like objects drops off as 1 over the square of the distance separating them. Why is that so? At least in principle, it seems that the force could drop off as simply 1 over the radius, or 1 over the radius cubed, or a myriad of other possibilities.

- There is a deep reason for this behavior. It boils down to the fact that the gravitational influence, or gravitational field, of an object such as a planet or a star spreads uniformly through space.

- Because we are exploring the idea that gravity might enter other dimensions of space, let's imagine what the law of gravity would be like if the world had a different number of dimensions than 3. If we represent the number of dimensions we have by the symbol N, the force between 2 objects will drop off as their separation to the N–1 power.

- This is a huge insight. The fact that Newton's law of universal gravitation says that the gravitational force between 2 objects drops off as the separation squared means that gravity exists in 3 dimensions.

- What are scientists thinking about when they talk about gravity maybe existing in more dimensions than the familiar 3? Surely they can't have overlooked something as simple as this, right?

- In 1919, Polish mathematician Theodor Kaluza proposed that there might be additional dimensions, and in 1926, Swedish physicist Oskar Klein suggested that the extra dimensions might be small. Klein's idea was in response to the criticism that we can obviously see that we live in a 3-dimensional world.

- Small dimensions are not infinite. This means that the small dimensions are cyclical. An example of a cyclical dimension might be a ring or the surface of a globe. If you walk on the surface of the Earth in a straight line for a long enough time, you will end up back where you started.

- The size at which we study objects can have a profound impact on the conclusions we draw. Measurements that can resolve objects that are about the size of extra dimensions and smaller can reveal truths about something as basic and central as the geometry of space-time itself. This points to a deep and core truth: that we need to take the idea of extra dimensions seriously. We shouldn't believe it, but we should be open it.

- Newton's law of gravity clearly favored a 3-dimensional universe. Newton's law has been tested very well in the realm of the solar system and in the calculations of orbital dynamics around the Earth. Furthermore, we can use sensitive instruments called gravitometers to study how gravity varies near the Earth to verify the predictions. But what about smaller distances?

- In 1797, British scientist Henry Cavendish used a torsion balance, which consisted of 2 lead balls, each weighing about a pound and a half, located on the ends of a wooden stick. The stick was suspended at its midpoint by a thread. Then, much heavier lead balls were brought near the smaller balls, and the stick was allowed to rotate under the influence of the gravitational force between the heavy and lighter lead weights.

- Using this apparatus, Cavendish was able to test Newton's law of gravity and even determine the gravitational constant, which he measured to a precision of 1%.

- Cavendish was able to verify Newton's theory down to distance scales of the size of about a centimeter. More recent measurements have improved on that. We have directly measured the behavior of the force of gravity for all distances above a hundredth of a millimeter and find that gravity falls off as 1 over the separation distance squared. This means that if extra dimensions exist, they are smaller than 10 micrometers.

- We might be able to make measurements at even shorter distances, but it's hard to imagine doing much better with torsion-balance technology. To do better, we need to use particle accelerators like the Large Hadron Collider. We've done that, and the experiments rule out extra dimensions bigger than the size of atoms or so. However, interpreting this data is tricky and depends on how many additional dimensions might exist. So, it's difficult to draw a firm conclusion on the basis of these investigations. But scientists are trying.

GRAVITONS

- Directly finding extra dimensions seems like it will be tricky, but there are some possibilities that are quite interesting. One topic involves gravitons, which are hypothetical particles that transmit the force due to gravity. They are massless, neutral particles with a quantum mechanical spin of 2. And given the weakness of gravity, there is no credible chance that we will ever be able to detect them. But in the context of extra dimensions, gravity isn't weak. So, what consequences will that have on the graviton?

- Suppose that a graviton was trapped in the small extra dimensions. This isn't so ridiculous. After all, the idea of extra dimensions was invented to explain the apparent weakness of gravity, and the solution was that gravity could enter these extra dimensions while the other particles couldn't. So, suppose that gravitons can enter the small dimensions.

- Now imagine what that would look like to us. After all, the extra dimensions are so small that we can't see they exist. That means that, according to us, the gravitons would look stationary. That's already weird. Normally, particles that have no mass must travel at the speed of light. And, of course, the gravitons do move at the speed of light—just not in dimensions we can see. So, we would draw other conclusions.

- A massless graviton can look massive if its motion is hidden in small and unobservable dimensions. In our familiar 3 dimensions, the object seems stationary. Because it has energy and no apparent motion, that means it appears to have mass.

- Thus, extra dimensions mean that we expect to be able to find massive gravitons. The particles will still be electrically neutral with a spin of 2, but they will appear to be massive because of the motion of the graviton in the hidden dimensions.

- This might seem sketchy, but this is pretty much the same thing as we saw on the topic of quantum chromodynamics. The proton and neutron have mass, even though they consist mostly of nearly massless up and down quarks. It is because the quarks are bound up in a small space. The mechanism here is different—the gravitons are bound in the small dimensions—but the outcome is the same.

- This is a huge prediction of extra-dimension theory: the existence of massive gravitons. Scientists call these Kaluza-Klein gravitons after Theodor Kaluza and Oskar Klein, who pioneered the idea nearly a century ago. The search for Kaluza-Klein gravitons is reputable.

BLACK HOLES

- If massive gravitons are kind of a peculiar idea, the next consequence of extra dimensions is pretty mind-blowing. Recall that extra dimensions were postulated as an explanation for why gravity is so weak. As the hypothesis goes, gravity is actually strong but appears weak because it leaks into the extra dimensions. And if we are able to create a way to look at things as small as the extra dimensions, we will see that gravity is actually strong.

- To look at small things, we need probes with very small wavelengths. These probes could be photons of light, or really any object. Due to the wave nature of matter and energy, high-energy objects have short wavelengths.

- That means that if we somehow generate particles with very high energies, they will have very short wavelengths and be able to probe the small extra dimensions. If we probe the extra dimensions, gravity will all of a sudden become strong.

- And that means that we will have concentrated a lot of energy in a small volume, and the result is that this is the condition for the creation of a black hole. Thus, if extra dimensions exist, and as we study matter under higher and higher energies, we will create tiny, subatomic, microscopic black holes.

- If extra dimensions do exist and we can make black holes, what would they look like? Because we've never seen one, the idea is speculative, but if black holes work in the way we think they do, black holes will disappear in an instant from Hawking radiation, named after British astrophysicist Stephen Hawking.

- Basically, his idea is that because the gravity is so strong in the vicinity of a black hole, some of the particles in the quantum foam will fall into the hole, while others will escape and carry away energy. Because energy and mass are equivalent, the loss of energy means that the microscopic black hole will lose mass and evaporate away in much less than the blink of an eye.

- Scientists are analyzing data at scientific facilities like the Large Hadron Collider to determine if they see the Hawking radiation of evaporating black holes. See the radiation and we have observed microscopic black holes. And if we see microscopic black holes, then we know that we're also seeing extra dimensions.

CONSEQUENCES

- The following is a graph of the way in which the unification of the forces seems to be going. Electroweak unification occurs at an energy of about 100 billion electron volts, while the electroweak and strong unification to a grand unified force occurs at 10^{15} billion electron volts. It is thought that gravity joins the party at a higher energy still, at 10^{19} billion electron volts.

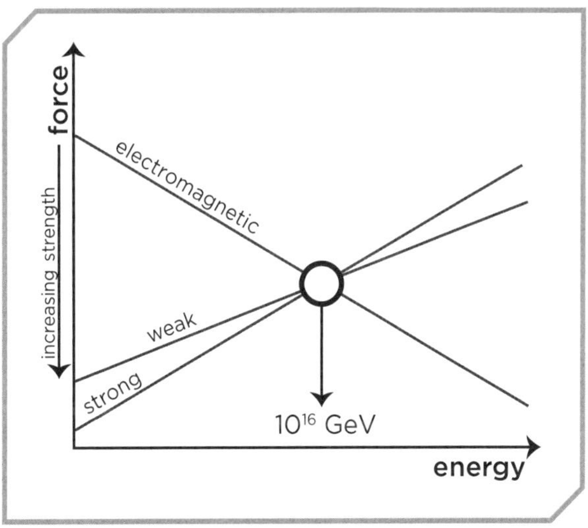

- But this picture changes radically if extra dimensions come into play, because when you get to a certain energy, you can begin to probe the extra dimensions, and at that point, all of our physics theories get tossed out the window. After all, the theories are built on equations written down using mathematics that rely on 3 infinite spatial dimensions and 1 of time.

- This would certainly change the unification graph, as gravity would suddenly become strong and interact on par with the other forces. The new graph would look like this:

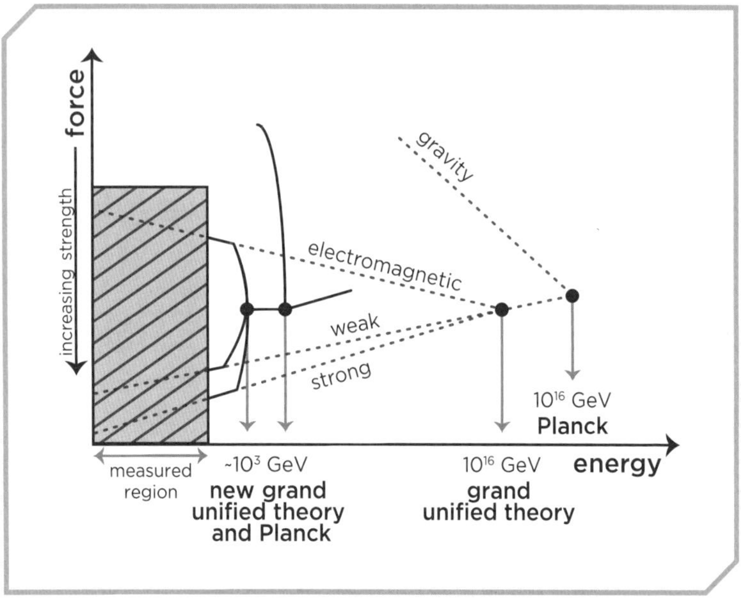

- The unification of the forces would occur at much lower energies. And this is a core question for our quest for unification. After all, the existence or nonexistence of extra dimensions will completely change the conversation about the unification of forces.

- After all, if no extra dimensions exist and the forces play out in 4 dimensions (meaning 3 of space and 1 of time), this ends up being a very different world than one in which more dimensions come into play. And don't forget that the reason that extra dimensions were proposed for this lecture was to account for the enormous weakness of the force of gravity.

- While this lecture focuses on gravity, the general idea of extra dimensions is broader than that. After all, in the previous lecture, the idea of extra dimensions also arose with superstrings.

- There is precisely zero evidence that extra dimensions actually exist, but they are not ruled out for sizes only a little smaller than the sizes we can investigate with current technology.

READINGS

Abbott, *Flatland.*
Krauss, *Hiding in the Mirror.*
Lincoln, *Understanding the Universe*, chap. 7.
Randall, *Warped Passages.*

QUESTIONS

1 How are parallel universes and extra dimensions different?

2 How can a massless graviton and extra dimensions combine in such a way so as to make the graviton appear to have mass?

22 BIG BANG AND INFLATION EXPLAIN OUR UNIVERSE

In addition to everything in the present, a theory of everything must also unify past and future, from our beginnings in the big bang all the way to the ultimate fate of the universe. There are questions of what came before the big bang and what exists outside our universe, but this lecture will focus on our universe by sketching out the big ideas of our universe as it currently is and revealing how we think the universe might have looked at the instant of creation—and then using physics to unify the origin of our universe with the universe we know today.

THE BIG BANG

- Using observations made by Edwin Hubble in 1929, we know that the universe is expanding, and thanks to astronomer Fred Hoyle, we now call it the big bang. According to the popular view of the theory, the universe was once concentrated into a point with zero volume—what scientists call a singularity—and it exploded.

- Singularities aren't real and mean that you've pushed a theory beyond the realm over which it applies. So, this singularity shouldn't be taken extremely seriously; it's just a reasonable approximation, and we can use it, even though we know it is wrong in detail.

- Most people have a mental image of the big bang as a grenade exploding, but this image is wrong, as it implies that there is a center to the universe and, if you had some kind of super-faster-than-light drive, that you could travel to the spot and stand on the location where the grenade went off.

- We know that the universe was born in a primordial fireball about 14 billion years ago. If we look to the cosmos in any direction, we find that galaxies are generally moving away from us, and the farther away they are, the faster away they move.

- Galaxies within about 10 million light-years are moving toward us, or with us, but galaxies far away are moving away from us with a certain velocity. Galaxies twice as far away are moving at double the velocity. Triple the distance and you get triple the velocity, and so on.

- This recessional velocity doesn't depend on the direction you are looking, and it's the same everywhere in the universe. It's not like a blast site, where some locations are closer than others; every place in the universe sees distant galaxies moving away in all directions.

SURPRISING UNIFORMITY

- In 1964, American physicists Arno Penzias and Robert Wilson discovered a hiss in the radio wave spectrum that was identified to be nothing less than the signature of the fireball of the big bang.

- If you think about an explosion in slow motion, it starts out white hot, and then, as it expands, the temperature drops and the color turns yellow, then orange, then red, and finally it cools to the point where it can't be seen. However, there are parts of the electromagnetic spectrum that also can't be seen—invisible colors. In the same way, while the temperatures of the big bang were once unfathomably high, emitting gamma waves from sources much hotter than those that create visible light, the universe has since expanded and cooled.

- In fact, the temperature of the universe—the temperature of the big bang—has now cooled all the way down to 2.7 Kelvin, or –455° Fahrenheit. An object with that temperature will emit radio waves, which is what Penzias and Wilson observed. The wavelength of these radio waves is about 2 millimeters, which is on the short end of the radio spectrum and overlaps with the microwave range. For that reason, this radio hiss is called the cosmic microwave background (CMB).

- There is one interesting feature of the CMB: It is very uniform. This uniformity is very puzzling. The CMB that we see coming from one direction was emitted essentially when the universe began, and it is just getting to us now. The same is true for the CMB coming from the opposite direction.

- That's a bit of a mystery, because the light from one side of the universe hasn't made it to the other side. So, in principle, the 2 sides of the universe were never in contact. The only way you'd expect to see that kind of uniformity was if all the points in the universe were once in the same place at the same time. Our best explanation for this is a rapid period of cosmic inflation, shortly after the big bang.

- There is one more curiosity about the CMB. While it is very uniform, in 1998, scientists observed that it had tiny variations—about 1 part in 100,000. This also needs an explanation, and we have an answer for that, too, and it might be the coolest explanation, as it ties together the smallest-imaginable phenomena with something as big as the entire visible universe.

- The uniformity of the CMB is but one aspect of the universe that is difficult to explain without a theory of everything. There are 2 others that are quite perplexing. The first is an amazing uniformity in the distribution of matter throughout the universe. That's an extraordinary observation that really needs explaining.

- But perhaps the most perplexing feature of the universe is why space itself has the shape it does. For example, space could be flat, curved, undulating, or any manner of distorted configurations. One of the consequences of Einstein's theory of general relativity is that space and time can be distorted, so the mathematical possibilities are endless.

- But space isn't all of those possibilities. In fact, to the best of our ability to measure it, space is flat—perfectly, mathematically, flat. We have been able to determine this by looking at those variations of temperature in the CMB radiation. We understand pretty much how they came to be, and we know how far away they are because we know how old the universe is.

- If space is distorted, it would distort the patterns we see in those spots. Yet when we look at them, they appear exactly as we would expect if space were perfectly flat on the largest scales. There are local deformations due to clusters of galaxies and local variations of matter, but when you ask the question of the overall structure of space, it is flat. And that's weird. Of all the possible configurations, flat is just one of them. Yet it's also somehow special. Why is space's configuration the special one?

INFLATION

- These handful of things are puzzling: the uniform expansion of the universe, the uniformity of the remnants of the primordial fireball, the uniform distribution of matter of the universe, and the flatness of space. These are all observations that we will have to explain with our theory of everything. So, what do we know?

- The short answer is that we don't know. But we have an idea that could bring it all together. This idea is called inflation and was first published in 1980 by American physicist Alan Guth. The basic idea is just an extension of the usual idea of the big bang.

- In inflation theory, the universe began and started to expand. Keep in mind that the big bang isn't what happened at time equals zero, but it's everything after time equals zero.

- The big bang began and the universe was expanding. At a time early in the process, the universe began to expand exponentially. The details are not known, but the universe expanded from a size far smaller than a proton to the size of a grapefruit or so in 10^{-32} seconds. And, with this one idea, Guth explained everything.

- When the universe was super tiny, it was all in contact and thus was homogeneous. The expansion pulled matter apart from one another, and the subsequent normal expansion of the big bang kept them apart, never to interact again. But the matter and energy were already imprinted with their initial homogeneity, so that explains the uniformity of the CMB and the distribution of matter.

- How inflation explains the flatness of space is even cooler. Suppose that before inflation space was actually distorted—for example, it had the shape of a sphere. No matter what the original shape of the universe was immediately after it began, inflation will always make it appear flat.

- What could have caused inflation? Basically, you need an energy source. The question is not answered, but there is at least one idea that is interesting. It comes from the concept of a phase transition, which is when matter or energy changes form somehow—such as when water turns to ice. In the transition, energy is extracted.

- Phase transitions are part of the theory. For example, if there is a grand unified force that includes both the strong and electroweak forces, when that force turns into 2 separate forces, that's a phase transition. In fact, the energy that is released during this transition is one candidate for inflation. It's not the only candidate, but it would be an incredibly elegant solution. It would combine both unification at the smallest-size scales with the very curious observed uniformity of the cosmos.

CURRENT THEORY

- Understanding the origins of our universe is a way to try to pull it all together. So, what do we know? We know of the standard model and we know of general relativity. We know how matter acts when it is stationary and how it acts as it approaches the speed of light. We know how matter works under temperatures near absolute zero—which is nearly −460° Fahrenheit—and as high as 4 trillion degrees Celsius. And we've explored the universe at scales as small as about 10^{-20} meters to the size of the universe itself, at 10^{27} meters.

- It is now possible to sketch out a timeline of the universe, from beginning to the present. Using particle accelerators, we understand the rules of the universe to energies approaching 10 trillion electron volts, which we think correspond to times after about 10^{-13} seconds after the big bang.

- Before about 10^{-13} seconds after the big bang, we speculate. In fact, that earlier realm reflects times and energies for which we need better experiments and better theories. Those theories could invoke extra dimensions, supersymmetry, superstrings, loop quantum gravity, or something that nobody has imagined yet.

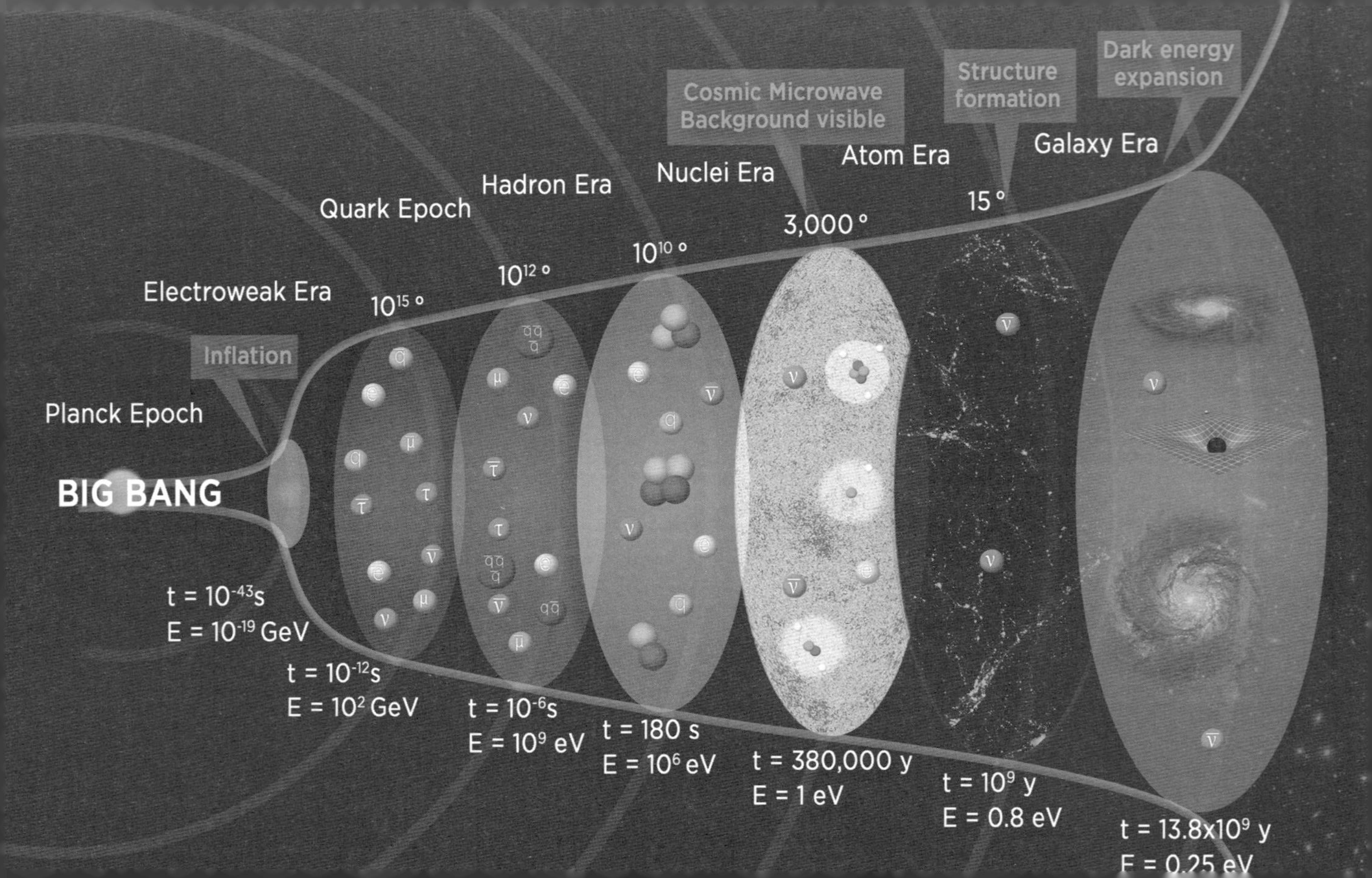
BIG BANG
Planck Epoch
Inflation
Electroweak Era
Quark Epoch
Hadron Era
Nuclei Era
Cosmic Microwave Background visible
Atom Era
Structure formation
Galaxy Era
Dark energy expansion
10^15 °
10^12 °
10^10 °
3,000 °
15 °
t = 10^-43 s
E = 10^-19 GeV
t = 10^-12 s
E = 10^2 GeV
t = 10^-6 s
E = 10^9 eV
t = 180 s
E = 10^6 eV
t = 380,000 y
E = 1 eV
t = 10^9 y
E = 0.8 eV
t = 13.8x10^9 y
E = 0.25 eV

- Indeed, if we ever want to understand how the universe came into existence and got where it is today, we must figure out a theory of everything. And if we figure out a theory of everything, we'll know how we came to be.

READINGS

Gilliland, *How to Build a Universe.*
Greene, *The Fabric of the Cosmos.*
Gribbin, *13.8.*
Singh, *Big Bang.*

QUESTIONS

1 Why are particle physics and cosmology so inextricably tied together?

2 The earliest time at which the laws of our universe could possibly apply is 10^{-43} seconds. We can recreate the conditions of the early universe as they existed at about 10^{-13} seconds after the big bang. What kinds of different phenomena might have existed between those 2 times? How much do you think that these new undiscovered phenomena might radically change our understanding of the early universe? Could it invalidate the big bang theory?

23 FREE PARAMETERS AND OTHER UNIVERSES

Does the universe have to be the way that it, is or could it be different? A good way to compare the universe we have with other possible universes is to look at 20 or so known but unexplained parameters of our universe. The core issue is that we do not have any theoretical basis for these parameters. If we had a theory of everything, we'd expect to be able to calculate the value of these parameters from first principles. Instead, we only know these parameters from experimental measurements, and we use these measured values and our current theories of the quantum and cosmic realms to perform other calculations. These are sometimes called free parameters. This lecture will explore how to explain these parameters.

OUR UTTERLY IMPROBABLE UNIVERSE

- We know that the standard model works quite well for the subatomic world and that general relativity handles gravity. We also believe it very likely that there is a better theory of everything that explains the connections between the quantum and cosmic worlds. A general presumption is that the free parameters of our current theories will one day be explained as originating from fewer parameters. Ideally, the theory will have zero free parameters, but at a minimum, it should be fewer.

- Using our 2 current theories, there are 20 parameters that we can't calculate from first principles and that we have to measure—19 of 20 of which come from the standard model. They are the masses of the 6 quarks and 3 charged leptons. We add the mass of the Higgs boson and the strength of the Higgs field. We then add 4 parameters that govern the details of how the quarks decay. We also need to understand known parameters for the strength of the strong, electromagnetic, and weak forces. There is a very technical parameter of quantum chromodynamics that describes the field when nothing is present. The last free parameter originates in general relativity and deals with the strength of gravity.

- Which of these parameters are really central to determining the structure of the universe? And if we change those parameters a little bit, does the universe look more or less the same, or is it completely different?

- The standard model describes all matter, including unstable particles, such as the top quark and the muon. But ordinary matter is made of just a few—protons, neutrons, and electrons, held together by the electromagnetic force.

- Given the quark composition of nucleons, that means that the most important parameters of the standard model as far as the day-to-day operations of the universe are concerned are the masses of the up and down quark, the mass of the electron, the strength of the electromagnetic force, and the force due to gravity. Those 5 parameters have a disproportionate impact on predicting our universe.

- The atomic level of matter is dominated by the mass of the electron and the strength of the electromagnetic field. Make the mass of the electron larger and atoms get smaller, and vice versa. Also, change the strength of the electric force and you will likewise change the size of atoms.
- However, these changes don't alter the basic structure of atoms. The atoms will be different in detail, but the big-picture aspects will remain unchanged. That is promising, inasmuch as if you have atoms, presumably you have chemistry—and, one assumes, life is possible.
- However, things get a bit more serious when the mass of the up and down quarks are considered. In our universe, the mass of the down quark is more than the mass of the up quark. Actually, it's more than the sum of the masses of the up quark and electron, which means a down quark can decay into an up quark plus an electron. Changing any of this would be a very big deal.

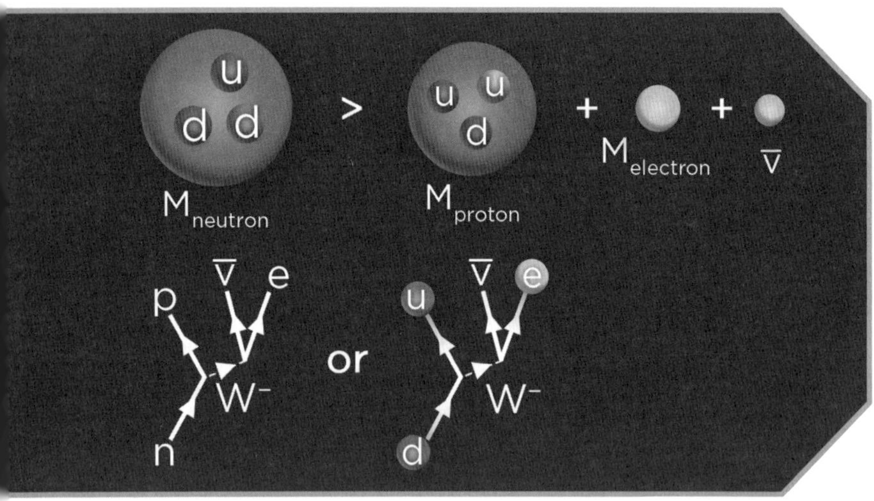

- The sum of the masses of the up and down quarks governs a lot of nuclear physics, which in turn governs if atoms like carbon can be made. It seems that the entire character of our universe is based very heavily on this very small difference in the masses of these 2 quarks. Change one of them and the existence of familiar matter would disappear.

- There are many similar examples of small changes that would radically change the look of the universe, but there is one that makes a bigger change: dark energy, which is an energy field that permeates the universe. It seems most likely that the energy field of the quantum foam might well be the origin of dark energy.

- However, there is the huge problem that the energy of quantum foam is 10^{120} times more than the dark energy that we measure. Something, therefore, has to counteract the huge quantum foam energy and bring it down to the dark energy that we see. And it can't cancel it exactly. And if dark energy were bigger than we see, then the universe's expansion would have been faster, making galaxies, stars, and all that impossible.

- It appears that at least some of the crucial parameters of the standard model and general relativity appear to be finely tuned to allow for our universe to exist.

- The exquisite fine-tuning of some important parameters leads some to use these seemingly impossible coincidences to postulate various religious or deistic ideas. But from a scientific point of view, these ideas are intrinsically untestable.

- There are a few testable ideas, and these ideas are connected to the question of what, if anything, happened before our universe came into existence.

EXPLAINING FINE-TUNING

- The first idea is the idea that a theory of everything will result in a single equation of the universe, which has a single solution. In this paradigm, the path that we have followed for centuries will continue, with the standard model eventually giving way to something better—perhaps involving supersymmetry or extra dimensions—and that theory giving way as more data is collected until, eventually, it is clear that the universe that we have simply had to be. The basic idea is that there is something in the very definition of a universe that mandates it be the one in which we live.

- It would be very interesting if that idea turns out to be true. But it also would be pretty peculiar that only one kind of universe can exist. It's also difficult to imagine that we will be able to answer this question anytime soon. But the biggest hurdle to this outcome arises from what we've learned about the universe since the advent of quantum mechanics nearly a century ago.

- Before around 1920, our understanding of the physics was governed by classical laws and with classical intuition. Cause was followed inexorably by effect.

- However, this all changed in the 1920s, when quantum mechanics was invented. Now there was a random element even in our universe. An atom could simultaneously be in multiple different configurations, and when we made a measurement, one of those configurations would be selected. Probabilities and wave functions governed the smallest scales.

- And when the universe began, it was much smaller. That means that it is almost certain that the right way to think about the universe as it began is quantum mechanically, governed by the laws of quantum physics.

- If we project what we know and think about the quantum world, that leads to a few ideas. One is that the universe just as it began had the possibility of existing in all possible universes—what scientists call a superposition of quantum configurations—and when it was formed, one of the various situations was selected. And that selection happened to be our universe. It could have been something very different, but it wasn't.

- That is a perfectly reasonable hypothesis, but one that is difficult to test and maybe undoable for a very long time.

- But there is another idea in quantum theory called the many worlds hypothesis, which says that every probabilistic quantum transition comes with a distinct universe. In this interpretation, all possible universes exist. Moreover, our experiences are just us wandering through the various possible timelines that could exist.

- If we project that basic mindset to the question of whether the universe has to be the way we think it is, this leads to an idea called the multiverse, which is short for multiple universes.

- There are many people who get turned off by the idea of multiverses. First, there is the definition problem. "Universe" is supposed to mean everything, so the idea of multiple everythings seems silly. But if the word "universe" is the visible universe governed by the laws of physics that we know, then there presumably could be other places that are governed by other laws. And those places might be different enough that it somehow seems appropriate to call them a different universe.

- The idea of multiple universes seems a bit outlandish and quite possibly untestable. But the hypothesis has gotten some traction in the theoretical community as a way to explain why the rules of our universe are the way they are. The hypothesis also has some possible bearing on the question of what happened before our universe began. And scientists are starting to think of ways that might make it possible to test the idea of a multiverse.

MULTIVERSES

- There are many ideas of what a multiverse might be. In particular, Swedish-American physicist Max Tegmark proposed 4 types of multiverses.
  1. The first one is based more or less on established physics. Essentially, the universe is infinite or at least incredibly large, and the laws and parameters of the universe vary over the vast volume of the cosmos. Some places can support matter and energy and life, while others can't. No place is special, but only in places that support complex matter can life form, and therefore we exist in such a volume.

  2. The second multiverse class is also tied crucially to the idea of inflation. Basically, in this idea, universes exist, and due to quantum mechanics, a small volume of a universe might spontaneously begin to inflate, thereby budding off from an existing universe. In this scenario, each universe might follow different laws, with the usual caveat that life can develop only in certain classes of these universes. In this model, universes have existed forever, but the moment at which a universe buds off from an existing universe, experiences inflation, and turns into a full-blown universe itself is the moment of that particular universe's creation.

  3. Tegmark's third class of multiverse is simply the multiworld interpretation of quantum mechanics. In this interpretation, early when the universe began, all of the possible configurations of physical laws and parameters existed in the quantum mechanical wave function of the universe. As the universe evolved, all of the possible configurations appeared in distinct universes. Thus, in our universe, the Higgs field appeared and gave mass to particles, while in other universes, the Higgs field may have never come into existence, or did so with a value of zero. In those universes, particles never gained mass, which resulted in a very different cosmos.

  4. In the fourth class of multiverses, Tegmark essentially argues that because abstract mathematics spans the entire realm of the possible, the multiverse is simply the mathematics.

- Some variants of the multiverse idea will probably remain forever untestable, but it may be that multiverses of Tegmark's second class might leave some observable signature on our universe. For example, if universes are floating around in some higher-dimensional space, then it is possible to imagine that these universes could collide.

- Most of the ideas that have been proposed for observable signatures of universes colliding with ours have to do with the cosmic microwave background radiation (CMB), which is the signature of the early universe. These ideas are not yet embraced by the scientific community and could well turn into nothing. But the real message is not that multiple universes are established, because they're not, but that the multiverse idea might lead to predictions that can be tested.

- In a sense, multiverse theory elevates the requirements of a theory of everything. It means that our theory must not only explain everything in our universe, but also it must explain everything in all possible universes.

READINGS

Clegg, *Before the Big Bang.*
Greene, *The Hidden Reality.*
Scientific American Editors, *Possibilities in Parallel.*
Susskind, *The Cosmic Landscape.*

QUESTIONS

1 Who is Professor Moriarty, and where was he last seen?

2 If a multiverse of type 2 is an accurate description of reality, then what are some basic ideas of what kinds of physics must describe the multiverse? Would they qualitatively differ from the basic ideas that describe our universe? This question should be approached as a mixture of philosophy and science.

24 TOWARD A FINAL THEORY OF EVERYTHING

The standard model and Einstein's theory of general relativity are still our best candidates of unified theories. But there are other phenomena that are not yet explained, and there are attempts to extend our theories with newer ideas, such as supersymmetry, extra dimensions, and quark and lepton compositeness. So, how exactly do we take the next step? How likely it is that we will make progress in the future, and where is progress most likely to be made? More pragmatically, what techniques will we use to try? These are the core questions for this lecture.

WHAT WE KNOW—AND DON'T KNOW

- In particle physics, there are many connections between various phenomena in the quantum world as well as informed speculative connections to the cosmos. The connections build a tree, but there are unknowns and linkages we don't understand at the moment.

- Both celestial and terrestrial gravity could be explained by Newton's theory of universal gravitation. That was back in the 1680s, so those connections are quite venerable. In the 1870s, we also learned how electricity, magnetism, chemistry, and light could all be explained by electromagnetism. These 2 unifications were the great classical unifications of physics.

- In the early 20th century, quantum mechanics unified atomic theory and Newtonian mechanics. Special relativity unified mass and energy. Then, quantum mechanics and special relativity were themselves combined with electromagnetism to make quantum electrodynamics (QED).

- Meanwhile, space and time were unified in a single geometry and combined with special relativity and Newton's theory of gravity to make Einstein's crowning achievement: the theory of general relativity.

- General relativity, unlike special relativity, was not unified with quantum mechanics. So, the quantum mechanics of charged particles, as described by quantum electrodynamics, and Einstein's general relativity were the 2 top-level theories.

- The turn into the 20th century also came with the discovery of radiation, the nucleus, and nuclear physics. A few decades later brought strongly interacting particles discovered in cosmic-ray measurements. The behavior of these laboratory and cosmic particles was eventually unified with quantum mechanics in a description known as quantum chromodynamics (QCD).

- Like QED, QCD again combines special relativity and quantum mechanics. We're not entirely sure how QED and QCD fit together; indeed, when we figure that out, we'll have made a big step forward toward a theory of everything.

- We also learned of the weak nuclear force, which we realized was tied to QED. This led to the unification now known as the electroweak force. But we also know that the pure electroweak force is incomplete. To have the electroweak force somehow turn into the 2 distinct forces called the weak force and QED, we needed a Band-Aid theory added to the whole situation called the Higgs field. We don't really know where the Higgs field comes from, but its existence was established in 2012, thereby confirming not only the electroweak unification, but also the entire standard model of particle physics.

- This diagram shows the connections that we know of. Basically, this is the standard model and general relativity. And this has been a very successful model. But it's not complete.

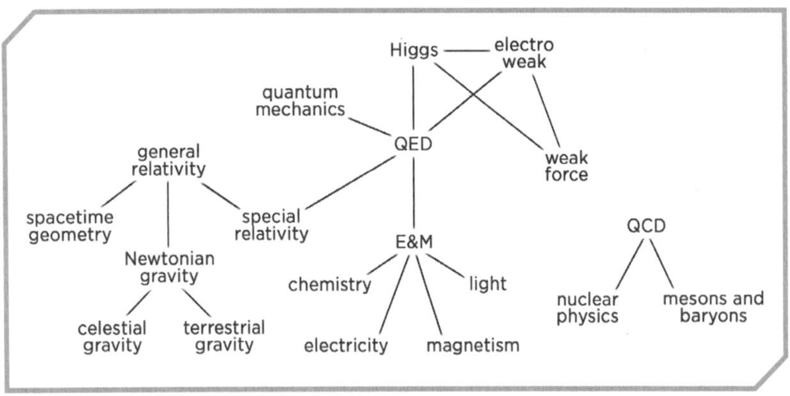

- For example, we can add other phenomena that we know exist but don't know where they fit into the picture, such as dark matter, dark energy, and the matter/antimatter asymmetry. There is the question of why there are carbon copies of the quarks and leptons, and the cause of cosmic inflation. And there is the very real question of what causes the Higgs field to exist.

- An important unification that we hope to make is to somehow combine quantum mechanics and general relativity and come up with a theory of quantum gravity. A second anticipated unification is that we can somehow connect the electroweak force and QCD to make a grand unified theory. If we're able to achieve both of those goals, we would then combine the grand unified theory and quantum gravity to make a theory of everything.

- But as great as that would be, it would still leave out a lot: Even if we are able to make all 3 of these unifications, we would be left with other known phenomena that don't have an understood place in this grand theoretical goal. And that's clearly not good. A theory of everything that doesn't incorporate everything is, by definition, not a theory of everything.

- Supersymmetry might find a role in taming mysteries in the Higgs theory. And the fact that dark matter has mass would probably connect it to the Higgs field. Supersymmetry might predict dark matter, and the fact that supersymmetry could answer both Higgs and dark-matter mysteries is one of the reasons that it is a popular idea in the theoretical community.

- Then, there is the possibility that dark energy is somehow related to inflation. After all, they both are energy fields that cause the universe to expand more rapidly; they are somehow the opposite of ordinary gravity. And they are energy fields that permeate the cosmos, just like the Higgs field.

- So, maybe dark energy, inflation, and the Higgs field are all connected somehow. And because dark energy is tied into gravity, it has to have a connection to general relativity. We don't know the details.

- Finally, there is always the very strong possibility that there are phenomena that we've not yet encountered that will result in more disconnected topics that we need to fit in somewhere.

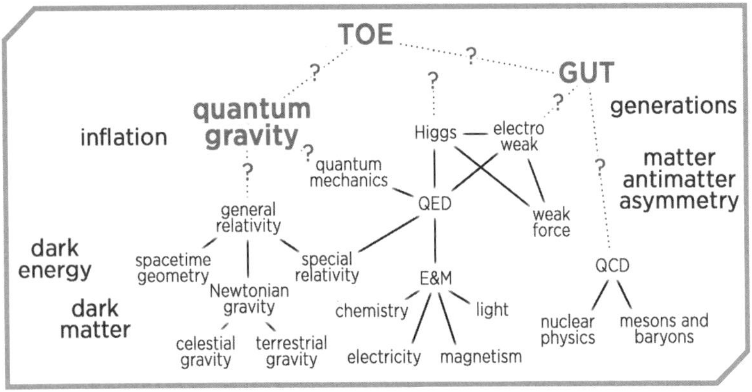

- Of course, a theory of everything should explain such things as the big bang and whatever was going on before our universe began. So, it's possible that even if we work out the interconnections of all of these myriad disconnected observations that the whole shebang is part of an even bigger theory that involves the multiverse or alternative universes, for example. The endgame for this whole endeavor is not very well known.

- Our current starting point is a single equation that contains everything covered in this course that is known to be true. It includes relativity, quantum mechanics, the standard model, and general relativity. The whole equation is the probability that a particle will transition from one state to another.

$$W = \overbrace{\int_{k<\Lambda} [Dg]\,[DA]\,[D\psi]\,[D\Phi]\, \exp}^{\text{quantum mechanics}} \Bigg\{ \overbrace{i \int d^4 x\, \sqrt{-g}}^{\text{spacetime}}\, \Bigg[\overbrace{\frac{m_p^2}{2} R}^{\text{gravity}}$$

$$\underbrace{-\frac{1}{4} F_{\mu\nu}^{a} F^{a\mu\nu}}_{\text{other forces}} + \underbrace{i\overline{\psi}^{i}\gamma^{\mu} D_{\mu}\psi^{i} + \left(\overline{\psi}_{L}^{i}\, V_{ij}\, \Phi \psi_{R}^{j} + \text{h.c.}\right)}_{\text{matter}} - \underbrace{\left| D_{\mu}\Phi \right|^{2} - V(\Phi)}_{\text{Higgs}} \Bigg] \Bigg\}$$

- We know that this isn't the final word, but it represents an impressive achievement. We want to replace it with something better. For example, this equation relies on 19 unknown parameters from the standard model and 1 from general relativity that we have to measure and add by hand. One goal of a better theory is to see how those unknowns are connected so that we can reduce the number of free parameters.

THE NEXT STEP

- How are we going to take the next step? One possibility is that the next Einstein will have an idea that pushes our understanding forward. But Einsteins are rare, and advances in understanding are often driven by data. So, what measurement techniques are scientists using to help make the next breakthrough?

- There are 2 measurement philosophies. One is to look at the many new theories proposed every week and see if the data agrees with the predictions. However, the limitation with that approach is that someone has to make a theory and a prediction. This is how the Higgs boson saga unfolded.

- The second method is to go looking for discrepancies. This approach means that we simply have to understand our current theory very well and then compare measurements to the predictions. Seeing an excess in some distribution will point us to the next new insight.

- What kinds of studies are we likely to do over the next couple of decades? There are 2 big classes, observational studies and experimental studies, with some smaller subclasses.

- Observational studies are an important approach that scientists will use to try to find dark matter. The approach has also been utilized for ideas such as proton decay and searches for magnetic monopoles. The gist of this technique is to put a detector somewhere appropriate on Earth and look for naturally occurring phenomena to be detected. This is called direct detection.

- Detectors can also be aimed at space, where we often have to use trace evidence to try to piece together what happened earlier. Here, scientists are still looking at naturally occurring phenomena, but the approach is more like a forensic investigator, piecing together clues that we have to understand what happened far away. This is called indirect detection. This method is also used to look for dark matter that is thought to be more plentiful near the center of galaxies, but it could also be used to investigate unknown phenomena that occur in the strong gravitational field near black holes.

- Looking into space is also important when studying dark energy. After all, that's how dark energy was observed in the first place. Scientists now look very deep into space, studying distant galaxies and supernovae, as well as how clusters of galaxies are organized. It is through astronomical studies that we'll figure out whether dark energy is a constant energy density or whether it is evolving over time. Knowing the character of dark energy will feed back into our grid of knowns and unknowns and maybe tell us where that piece fits into the puzzle.

- Besides observational approaches (direct and indirect), we can do experiments on Earth to try to make new phenomena. The basic idea uses particle colliders to smash 2 known kinds of particles together to concentrate a lot of energy in one spot to see if we can make new particles. Basically, this just uses Einstein's $E = mc^2$ equation, which shows how energy and mass are related.

- Another feature of using high-energy particles to study the world hinges on Louis de Broglie's observation that particles are also waves. The wavelength of a particle gets smaller and smaller as its energy increases. And short wavelengths mean that you can see smaller things: X-rays have wavelengths that are short enough to investigate the crystal structure of molecules, and electron microscopes have made it possible to image individual atoms. So, very high-energy collisions might reveal a hypothetical new level, with new kinds of particles smaller than quarks.

- What kinds of particle accelerators are likely to make important contributions over the next few decades? For probably 20 years, the Large Hadron Collider (LHC) will be the highest-energy particle accelerator in the world. It was designed in part to discover the Higgs boson, which happened in 2012, but it was really designed to be much more. It was built to explore energies that haven't been common in the universe since it was a tenth of a trillionth of a second old.

- The LHC is certainly one of the most likely scientific facilities to teach us something new and fundamental about the laws of the universe. Data from the LHC might be used to develop better theories that allow us to figure out where one of the unattached mysteries fits into our overall diagram.

- But this is not the only way to do experiments. The predictions of modern theories are all about probability. And that means that rare—even classically impossible—things can happen. In particle physics, if a particular kind of interaction is incredibly rare and very unlikely to occur in the collision of any specific pair of particles, then you improve your odds greatly if you simply collide a huge number of particles. This approach is a perfectly credible one. The technical term is to have huge luminosity.

- Trying to make extremely high-luminosity beams of particles is the approach taken in the U.S. laboratory system. Fermi National Accelerator Laboratory, known as Fermilab, has elected to go this route. It will allow unprecedented studies of the interactions of neutrinos, which have surprised us many times in the past. It also allows incredibly precise studies of quantum electrodynamics through the study of the spin of muons, which will offer a precision of parts per trillion.

- Both of these techniques have to be explored if we hope to make observations that will allow us to figure out what is the next step. And these aren't the only ones. There are also scientists who are trying to make ultraprecise measurements using detectors that simply study atoms. If there are other forces or interactions that we haven't discovered, then maybe such interactions affect things like the energy levels of atoms—and very precise measurements will reveal their presence.

- Scientists are exploring ideas of how to make accelerators that are even bigger and better than existing ones. They may never be built, but discussions are always underway about how to build a particle accelerator that is 10 times more powerful than the LHC.

READINGS

Carroll, *The Big Picture.*
Hawking, *The Theory of Everything.*
Rovelli, *Seven Brief Lessons on Physics.*

QUESTIONS

1. Which of the techniques discussed in this lecture do you think are more promising for future new paradigm-changing discoveries?

2. What do you think is a reasonable estimate for the amount of time it will take for mankind to devise a theory of everything?

BIBLIOGRAPHY

Abbott, Edwin. *Flatland: A Romance of Many Dimensions*. London: Dover, 1992. A fictional and very accessible story about multiple dimensions.

Bennett, Jeffrey. *What Is Relativity?: An Intuitive Introduction to Einstein's Ideas, and Why They Matter*. New York: Columbia University Press, 2014. A book that stresses the intuitive aspects of relativity.

Butterworth, Jon. *Most Wanted Particle: The Inside Story of the Hunt for the Higgs, the Heart of the Future of Physics*. New York: The Experiment, 2015. Butterworth is both a practicing physicist and a frequent writer for *The Guardian*. He tells the inside story of the discovery of the Higgs boson.

Carroll, Sean. *Spacetime and Geometry: An Introduction to General Relativity*. New York: Pearson, 2003. This book is a textbook and not light reading. However, for a dedicated reader wishing to delve deeply into the subject, it is very good.

———. *The Big Picture: On the Origins of Life, Meaning and the Universe Itself.* New York: Dutton, 2016. This book is much more philosophical than most recommended books. It weaves together what we know and what we can know. It is not recommended for a fact-centric person.

———. *The Particle at the End of the Universe: How the Hunt for the Higgs Boson Leads Us to the Edge of a New World*. New York: Dutton, 2013. This book is a story of the discovery of the Higgs boson, written by a theoretical physicist who was not directly involved in the discovery. It focuses on the manner in which the Higgs boson fits into our contemporary understanding of the universe and not on the mechanics of how it was found.

Clegg, Brian. *Before the Big Bang: The Prehistory of the Universe.* New York: St. Martin's Griffin, 2011. This book is an ambitious attempt to try to figure out what might have happened before the big bang occurred.

Close, Frank. *Antimatter.* Oxford: Oxford University, 2010. This small book is focused on the subject of antimatter, including the very pressing mystery of why there is so little of it to be found in our universe. This is one of the big unsolved mysteries.

———. *Neutrino.* Oxford: Oxford University, 2012. This small book focuses exclusively on the enigmatic neutrino, which has surprised us many times in the last 80 years and probably will again.

———. *The Infinity Puzzle: Quantum Field Theory and the Hunt for an Orderly Universe.* New York: Basic, 2013. This book is unusual in that it describes the inner mechanics of quantum field theory in an accessible way. Most books gloss over this facet of physics.

———. *The New Cosmic Onion: Quarks and the Nature of the Universe.* Boca Raton, FL: CRC Press, 2006. This book covers the history of particle physics and the standard model. It incorporates more mathematics than most popularizations or semi-popularizations and gives the reader a taste of what it is like to calculate simple quantities.

Collier, Peter. *A Most Incomprehensible Thing: Notes towards a Very Gentle Introduction to the Mathematics of Relativity.* Unknown: Incomprehensible Books, 2014. This book attempts to introduce the basic mathematics of relativity to a person with a moderate mathematical background. It focuses on the mathematical introduction and not the interpretation.

Einstein, Albert. *Relativity: The Special and the General Theory.* Translated by Robert Lawson. New York: LG Classics, 2013. This book was written by Einstein and is a little difficult. Later authors have done a better job of simplifying the subject, but it is delightful to read about the subject in Einstein's own (translated) words.

Farmelo, Graham. *The Strangest Man: The Hidden Life of Paul Dirac, Mystic of the Atom*. New York: Basic Books, 2011. Paul Dirac was one of the key architects of relativistic quantum mechanics and an eccentric fellow. This book blends the history of the man and the physics that he developed.

Feynman, Richard P. *QED: The Strange Theory of Light and Matter*. Princeton, NJ: Princeton Science Library, 2014. Feynman is an excellent writer and one of the architects of quantum electrodynamics. He tackles this difficult subject with a minimum of math and highlights conceptual aspects of the first quantum field theory.

Forbes, Nancy, and Basil Mahon. *Faraday, Maxwell and the Electromagnetic Field: How Two Men Revolutionized Physics*. New York: Prometheus Books, 2014. Maxwell and Faraday were 2 of the giants in the effort to unify electromagnetism. This book sets their physical discoveries in a historical context.

Gamow, George. *Gravity*. Garden City, NY: Dover, 2003. Gamow's discussion of gravity covers the material in a very accessible way.

———. *Mr. Tompkins in Paperback*. Cambridge: Cambridge University Press, 1983. Gamow's Mr. Tompkins books are legendary. They are fictionalized accounts of a regular guy encountering quantum and relativistic phenomena in a world in which they are visible in day-to-day to life. A must-read.

———. *Thirty Years That Shook Physics: The Story of Quantum Theory*. Garden City, NJ: Dover Publications, 1985. Gamow covers both the history and the physics of the quantum revolution.

Gardner, Martin. *Relativity Simply Explained*. Garden City, NJ: Dover, 1997. Gardner was a frequent columnist for *Scientific American* and an expert writer. This book does an excellent job of describing relativity in a simple way.

Gates, Evalyn. *Einstein's Telescope: The Hunt for Dark Matter and Dark Energy in the Universe*. New York: W. W. Norton, 2010. This book covers the hunt for both dark matter and dark energy.

Gibilisco, Stan. *Electricity Demystified*. New York: McGraw-Hill Education, 2012. This is one of the Demystified series, which is akin to the "X *for Dummies*" books. It explains electricity and magnetism very simply and with accessible mathematics.

Gilliland, Ben. *How to Build a Universe: From the Big Bang to the End of the Universe*. London: Sterling, 2015. This book is written about the birth and death of the universe in the manner of a journalist, which is what Gilliland is by training.

Gilmore, Robert. *Alice in Quantumland*. New York: Copernicus, 1995. An interesting admixture of fiction and physics in a blending of *Alice in Wonderland* and quantum mechanics. It is similar to Abbott's *Flatland*.

Giudice, Gian Francesco. *A Zeptospace Odyssey: A Journey into the Physics of the LHC*. Oxford: Oxford University Press, 2010. One of the early books written about the research program at the Large Hadron Collider (LHC). It came out prior to the start of operations and looks forward to the discoveries that the LHC might make.

Greene, Brian. *The Elegant Universe: Superstrings, Hidden Dimensions, and the Quest for the Ultimate Theory*. New York: W. W. Norton, 2010. Greene's book is the seminal popularization about superstrings.

———. *The Fabric of the Cosmos: Space, Time and the Texture of Reality*. New York: Vintage, 2005. Greene's book covers the structure and shape of the universe. This book is more about relativity and is a very nice read.

———. *The Hidden Reality: Parallel Universes and the Deep Laws of the Cosmos*. New York: Vintage, 2011. This book talks about questions of whether the universe has to be the way it is and invokes speculative ideas, such as parallel universes, as a method to explain the very curious question of why the universe seems to be so tuned to permit humans to exist.

Gribbin, John. *13.8: The Quest to Find the True Age of the Universe and the Theory of Everything*. New Haven, CT: Yale University, 2016. Figuring out the age of the universe is very tricky. In this book, Gribbin reviews the experimental evidence that has led astronomers to determine that the age of the universe is 13.8 billion years old. The book also brings together the subatomic and cosmic worlds.

———. *In Search of Schrodinger's Cat: Quantum Physics and Reality*. New York: Bantam Books 1984. A book that covers the counterintuitive aspects of quantum mechanics.

Griffiths, David. *Introduction to Electromagnetism*. 4th ed. Essex, UK: Pearson, 2012. A college-level textbook on electromagnetism. It is very well written but assumes a fairly sophisticated student.

———. *Introduction to Elementary Particle Physics*. 2nd ed. Weinheim, Germany: Wiley-VCH, 2008. A college-level textbook on elementary particle physics. It is very well written but assumes a fairly sophisticated student.

———. *Introduction to Quantum Mechanics*. 2nd ed. Cambridge, UK: Pearson Prentice Hall 2014. A college-level textbook on quantum mechanics. It is very well written but assumes a fairly sophisticated student.

Hawking, Steven. *The Theory of Everything*. Mumbai, India: Jaico Publishing, 2007. Hawking's musings on what a theory of everything might entail.

Hoddeson, Lillian, Laurie Brown, Michael Riordan, and Max Dresden. *The Rise of the Standard Model: Particle Physics in the 1960s and 1970s*. Cambridge: Cambridge University Press, 1997. A book detailing the developmental stages of the standard model, which was from about 1960 to about 1980.

Hooper, Dan. *Dark Cosmos: In Search of Our Universe's Missing Mass and Energy*. New York: Harper Perennial, 2007. This book is an excellent introduction to dark matter and dark energy for a reader who knows very little about it. It is a first introduction and not as suited for a more sophisticated reader.

———. *Nature's Blueprint*. New York: Smithsonian, 2008. Hooper's book on supersymmetry covers the reasons that theoretical physicists are so fascinated by this unproven hypothesis.

Jayawardhana, Ray. *Neutrino Hunters: The Thrilling Chase for a Ghostly Particle to Unlock the Secrets of the Universe*. New York: Scientific American/Farrar, Straus and Giroux, 2015. An excellent coverage of the modern effort of scientists to use beams of neutrinos to unlock the mysteries of the universe. The book also includes discussion of studies of neutrinos from space.

Kalman, Calvin, and Ian D. Souza. *Preons: Models of Leptons, Quarks and Gauge Bosons as Composite Objects*. Singapore: World Scientific, 1992. This is a fairly technical book, but it is included because there are very few books that talk about the possibility of quark and lepton constituents.

Kane, Gordon. *Supersymmetry: Squarks, Photinos, and the Unveiling of the Ultimate Laws of Nature*. New York: Perseus, 2000. Kane has been a passionate proponent of supersymmetry for many decades. He writes well but a little uncritically on the subject.

———. *Supersymmetry: Unveiling the Ultimate Laws of Nature*. New York: Basic, 2001. Kane has been a passionate proponent of supersymmetry for many decades. He writes well but a little uncritically on the subject.

———. *The Particle Garden*. New York: Helix Books, 1996. This is a short book about the Standard model that is slightly dated and leans more toward the theoretical aspects but is very well written and highly accessible.

Krauss, Lawrence. *Hiding in the Mirror: The Mysterious Allure of Extra Dimensions, from Plato to String Theory and Beyond.* New York: Viking Adult, 2005. A historical treatment of multiple dimensions with an occasional nod to pop culture.

Labelle, Patrick. *Supersymmetry Demystified.* New York: McGraw-Hill Education, 2010. Part of the Demystified series (similar to the "X *for Dummies*" books) on the subject of supersymmetry.

Lederman, Leon, and Chris Hill. *Beyond the God Particle.* New York: Prometheus, 2013. This book talks about what the theoretical and experimental landscape will be now that the Higgs boson has been discovered. It's a good primer for the near term in terms of what to expect next.

———. *Symmetry and the Beautiful Universe.* New York: Prometheus, 2004. This book is a lovely introduction to the role of symmetry in modern particle theories.

Lederman, Leon, and Dick Teresi. *The God Particle: If the Universe Is the Answer, What Is the Question?.* New York: Mariner, 2006. Leon Lederman is responsible for the use of the term "the god particle" to mean the Higgs boson. Ironically, this book is more about the history of particle physics, including Lederman's role in key discoveries of the weak force, than it is about the Higgs boson itself.

Lincoln, Don. "The Inner Life of Quarks." *Scientific American* 307 (November 2012): 22–29. This article is one of the few treatments of the idea of quark and lepton substructure. It is an old idea, but not one that has been popularized lately.

———. *The Large Hadron Collider: The Extraordinary Story of the Higgs Boson and Other Things That Will Blow Your Mind.* Baltimore, MD: Johns Hopkins, 2014. This book talks about the standard model, including the discovery of the Higgs boson. After an in-depth discussion about the Large Hadron Collider (LHC), the book covers the subjects that LHC scientists are focusing on for future possible near-term discoveries.

———. *Understanding the Universe: From Quarks to the Cosmos.* rev. ed. Singapore: World Scientific, 2012. This book offers an encyclopedic look at particle physics and its connection to cosmology. It has an extensive historical section and covers pressing unsolved mysteries.

Livio, Mario. *The Equation That Couldn't Be Solved: How Mathematical Genius Discovered the Language of Symmetry.* New York: Simon and Schuster, 2005. This book is built around the unsolvable so-called quintic equation and how investigations into the equation led to group theory and an appreciation of mathematical symmetries. It is not math-heavy and has a substantial dollop of human-interest stories.

Mahon, Basil. *The Man Who Changed Everything: The Life of James Clerk Maxwell.* New York: Wiley, 2004. Mostly a biography, but with some physics. It is good at setting historical context of his amazing contributions to electromagnetism.

Nicolson, Iain. *Dark Side of the Universe: Dark Matter, Dark Energy, and the Fate of the Cosmos.* Baltimore, MD: Johns Hopkins, 2007. A good and accessible broad survey of dark matter and dark energy.

Pais, Abraham, Maurice Jacob, David I. Olive, and Michael F. Atiyah. *Paul Dirac: The Man and His Work.* Cambridge: Cambridge University Press, 1998. Pais was both a physicist and a science historian who personally knew Dirac. This is a great biography of a great man.

Panek, Richard. *The 4 Percent Universe: Dark Matter, Dark Energy, and the Race to Discover the Rest of Reality.* New York: Mariner, 2011. A good general-purpose survey of dark matter and dark energy.

Pask, Colin. *Magnificent* Principia*: Exploring Isaac Newton's Masterpiece.* New York: Prometheus Books, 2013. Newton's *Principia* is one of the most seminal scientific treatises ever written, but it's not easy reading. Pask guides you through the *Principia* in a way that makes it easier to understand and allows you insight into the great man's mind.

Randall, Lisa. *Dark Matter and the Dinosaurs: The Astounding Interconnectedness of the Universe*. New York: Ecco, 2015. Randall is one of the architects of the idea of complex dark matter, which is dark matter that forms complicated structures but doesn't interact with ordinary matter. She's an excellent writer, and the excursion into linking dark matter and the death of dinosaurs is perhaps an excusable bit of fluff.

———. *Higgs Discovery: The Power of Empty Space*. New York: Ecco, 2013. Randall wrote this book quickly after the discovery of the Higgs boson. Accordingly, it is short and fresh, although it is a little thin due to the speed at which it was prepared.

———. *Warped Passages: Unraveling the Mysteries of the Universe's Hidden Dimensions*. New York: Harper Perennial, 2006. A delightful and accessible book on the idea of extra dimensions by one of the core minds behind the idea. It is well worth your time.

Rovelli, Carlo. *Seven Brief Lessons on Physics*. New York: Riverhead Books, 2016. A short book that covers some important ideas in physics: Quantum mechanics, general relativity, quantum gravity, and black holes are among the most important topics for this course.

Sample, Ian. *Massive: The Missing Particle That Sparked the Greatest Hunt in Science*. rev. ed. New York: Basic, 2013. This extraordinary book gives the human story behind the creation of the theory of the Higgs field. It is a history and not a physics book, but it is a truly delightful story.

Schumm, Bruce. *Deep Down Things: The Breathtaking Beauty of Particle Physics*. Baltimore, MD: Johns Hopkins, 2004. This is an unusual, perhaps unique, book. It talks about the standard model in terms of the symmetries underlying the ideas and much less about the nuts and bolts of quarks and leptons. It is a theoretical book, written by an experimentalist, and without the daunting math.

Schweber, Silvan. *QED and the Men Who Made It*. Princeton, NJ: Princeton University Press, 1994. The story of quantum electrodynamics would not be complete without an understanding of the historical context in which it was developed. This book blends both physics and history.

Scientific American Editors. *Possibilities in Parallel: Seeking the Multiverse*. New York: Scientific American, 2013. A series of essays guest authored by editors for *Scientific American*. It is a short book that is rich with key ideas.

Singh, Simon. *Big Bang: The Origin of the Universe*. New York: Harper Perennial, 2005. A good description of the big bang theory.

Smolin, Lee. *The Trouble with Physics: The Rise of String Theory, The Fall of a Science, and What Comes Next*. New York: Mariner, 2007. String theory is notorious for being unable to make testable predictions. Smolin's book talks about string theory and its weaknesses as a scientific theory.

———. *Three Roads to Quantum Gravity*. New York: Basic, 2002. There are several ideas about how the quantum realm and gravity will be connected. Smolin's book covers superstrings, quantum gravity, the holographic principle, and more.

Stewart, Ian. *Why Beauty Is Truth: The History of Symmetry*. New York: Basic, 2007. Symmetry really is key in understanding the mathematics of particle physics theories. This book explores this key thought.

Styer, Daniel. *Relativity for the Questioning Mind*. Baltimore, MD: Johns Hopkins 2011. An extraordinary book on relativity, perhaps the easiest and clearest explanation that exists. Notable are the illustrations, which show how different observers experience the same thing.

Susskind, Leonard. *The Cosmic Landscape: String Theory and the Illusion of Intelligent Design*. New York: Back Bay 2006. This book, written by one of the architects of superstring theory, addresses the question of why we exist at all. It explores the possible reasons for why the universe seems so well-tuned to allow humans to exist.

Susskind, Leonard, and Art Friedman. *Quantum Mechanics: The Theoretical Minimum*. New York: Basic Books, 2015. This delightful book is more challenging than most. It covers what Susskind claims is the minimum amount of theoretical and mathematical ideas needed to understand physics theories.

Veltman, Martinius. *Facts and Mysteries in Elementary Particle Physics*. Singapore: World Scientific, 2003. Veltman's book is a simple introduction to the standard model. The book's weakness is that it was written before the Higgs boson was found, but its strength are the insider stories that the Nobel Prize winner sprinkles throughout the pages.

Wilczek, Frank. "QCD Made Simple." *Physics Today* 53, no. 8 (August 2000): 20. This article offers a clear and lovely description of quantum chromodynamics by one of the theory's architects.

Woit, Peter. *Not Even Wrong: The Failure of String Theory and the Search for Unity in Physical Law*. New York: Basic, 2007. String theory has been the darling of the theoretical community for decades, but its inability to make testable predictions has caused it to lose some of its luster. Woit's book focuses on the weaknesses of this once-popular theory.

Wolfson, Richard. *Simply Einstein: Relativity Demystified*. New York: W. W. Norton, 2003. This book is one of the Demystified series, which is similar to the "X *for Dummies*" books. It offers a nuts-and-bolts description of relativity.

IMAGE CREDITS

PAGE NUMBER

5 .... © GeorgiosArt/iStock/Thinkstock.
6 .... Serge Lachinov/Wikimedia Commons/Public Domain.
8 .... The Teaching Company Collection.
13 .... Library of Congress Prints and Photographs Division, LC-USZ62-60242.
15 .... © Dorling Kindersley/Thinkstock.
16 .... © Photos.com/Thinkstock.
17 .... © Photos.com/Thinkstock.
17 .... Science Museum, London/Wellcome Images/CC BY-4.0.
19 .... © Photos.com/Thinkstock.
20 .... © Photos.com/Thinkstock.
25 .... Images from the History of Medicine (NLM).
27 .... © PeterHermesFurian/iStock/Thinkstock.
29 .... Library of Congress Prints and Photographs Division, LC-DIG-ggbain-03392.
30 .... The Teaching Company Collection.
32 .... The Teaching Company Collection.
42 .... Kelly - University of Toronto/Internet Archive.
51 .... The Teaching Company Collection.
52 .... The Teaching Company Collection.
53 .... The Teaching Company Collection.
57 .... RockMagnetist/Wikimedia Commons/Public Domain.
58 .... Savannah River Site/flickr/CC BY-2.0.
61 .... Photo courtesy of the U.S. Department of Energy.
62 .... Acc. 90-105 - Science Service, Records, 1920s-1970s, Smithsonian Institution Archivess
62 .... The Teaching Company Collection.
63 .... The Teaching Company Collection.

63 .... The Teaching Company Collection.
68 .... National Archives and Records Administration.
69 .... CERN.
72 .... The Teaching Company Collection.
72 .... The Teaching Company Collection.
72 .... The Teaching Company Collection.
73 .... The Teaching Company Collection.
78 .... Larry D. Moore/Wikimedia Commons/CC BY-SA 3.0.
78 .... Molendijk, Bart/Anefo/Wikimedia Commons/CC BY-SA 3.0 nl.
79 .... Bengt Nyman/Wikimedia Commons/CC BY-2.0.
86 .... The Teaching Company Collection.
89 .... World Economic Forum/Wikimedia Commons/CC BY-SA 2.0.
89 .... 和平奮救地球/Wikimedia Commons/CC BY-SA 4.0.
95 .... The Teaching Company Collection.
99 .... The Teaching Company Collection.
108 .... The Teaching Company Collection.
114 .... The Teaching Company Collection.
114 .... The Teaching Company Collection.
115 .... The Teaching Company Collection.
115 .... The Teaching Company Collection.
142 .... The Teaching Company Collection.
154 .... © Morrison1977/iStock/Thinkstock.
161 .... © Paul Fleet/iStock/Thinkstock.
165 .... J. Burrus/NIST.
171 .... The Teaching Company Collection.
182 .... The Teaching Company Collection.
201 .... The Teaching Company Collection.
202 .... The Teaching Company Collection.
210 .... The Teaching Company Collection.
214 .... The Teaching Company Collection.

NOTES